AF400886

CONTRAST

PHOTOGRAPHY ON THE LONDON UNDERGROUND

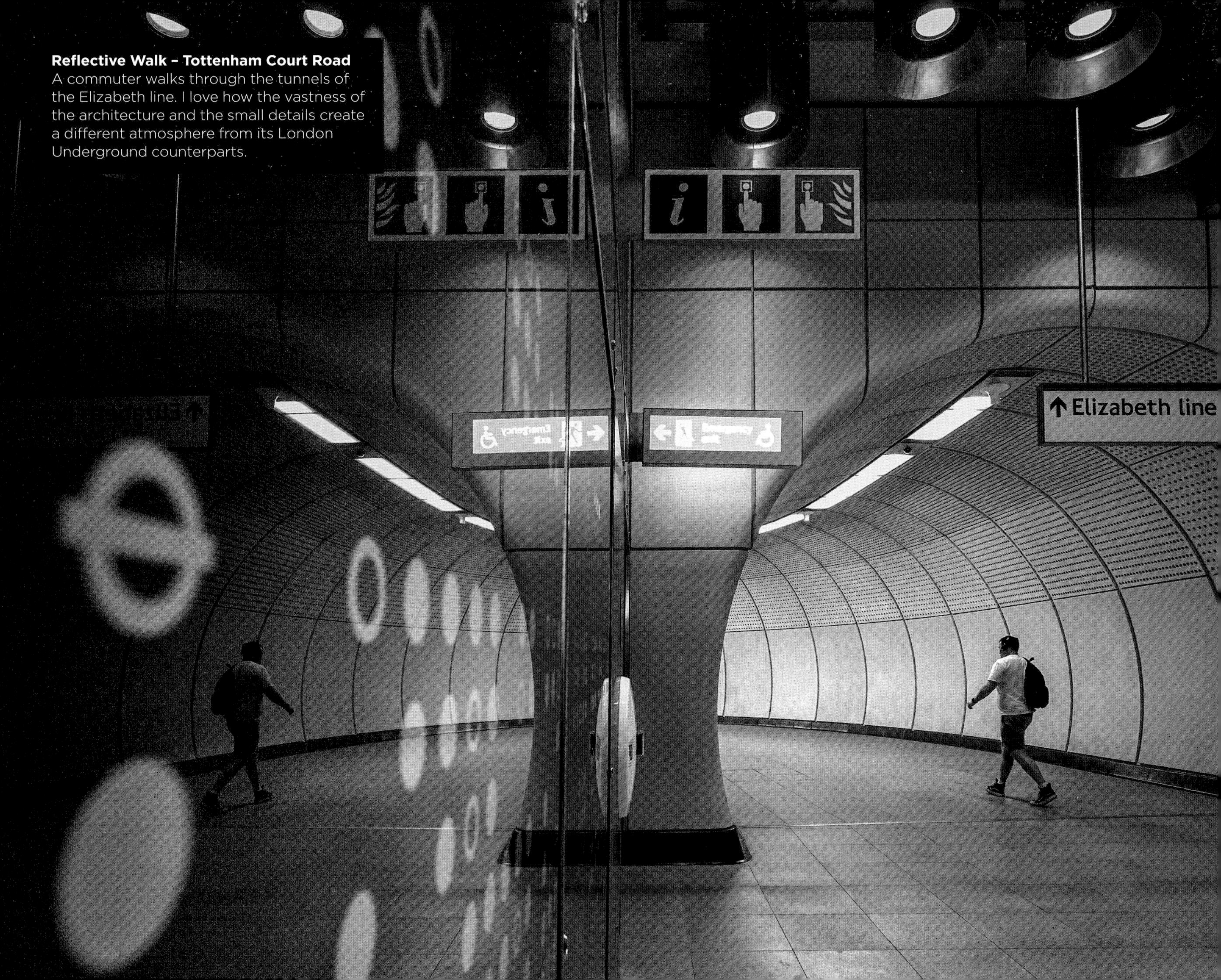

Reflective Walk – Tottenham Court Road
A commuter walks through the tunnels of the Elizabeth line. I love how the vastness of the architecture and the small details create a different atmosphere from its London Underground counterparts.

CONTRAST

PHOTOGRAPHY ON THE LONDON UNDERGROUND

THE TUBE MAPPER PROJECT

LUKE AGBAIMONI

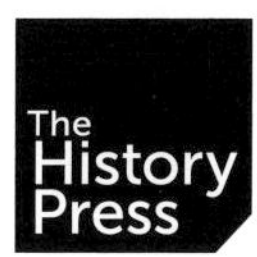

For Ellie, River & Fox

First published 2024

The History Press
97 St George's Place, Cheltenham,
Gloucestershire, GL50 3QB
www.thehistorypress.co.uk

British Library Cataloguing in Publication Data.
A catalogue record for this book is available from the British Library.

ISBN 978 1 80399 764 3

Typesetting and origination by The History Press
Printed and bound in India by Thomson Press India Ltd

CONTENTS

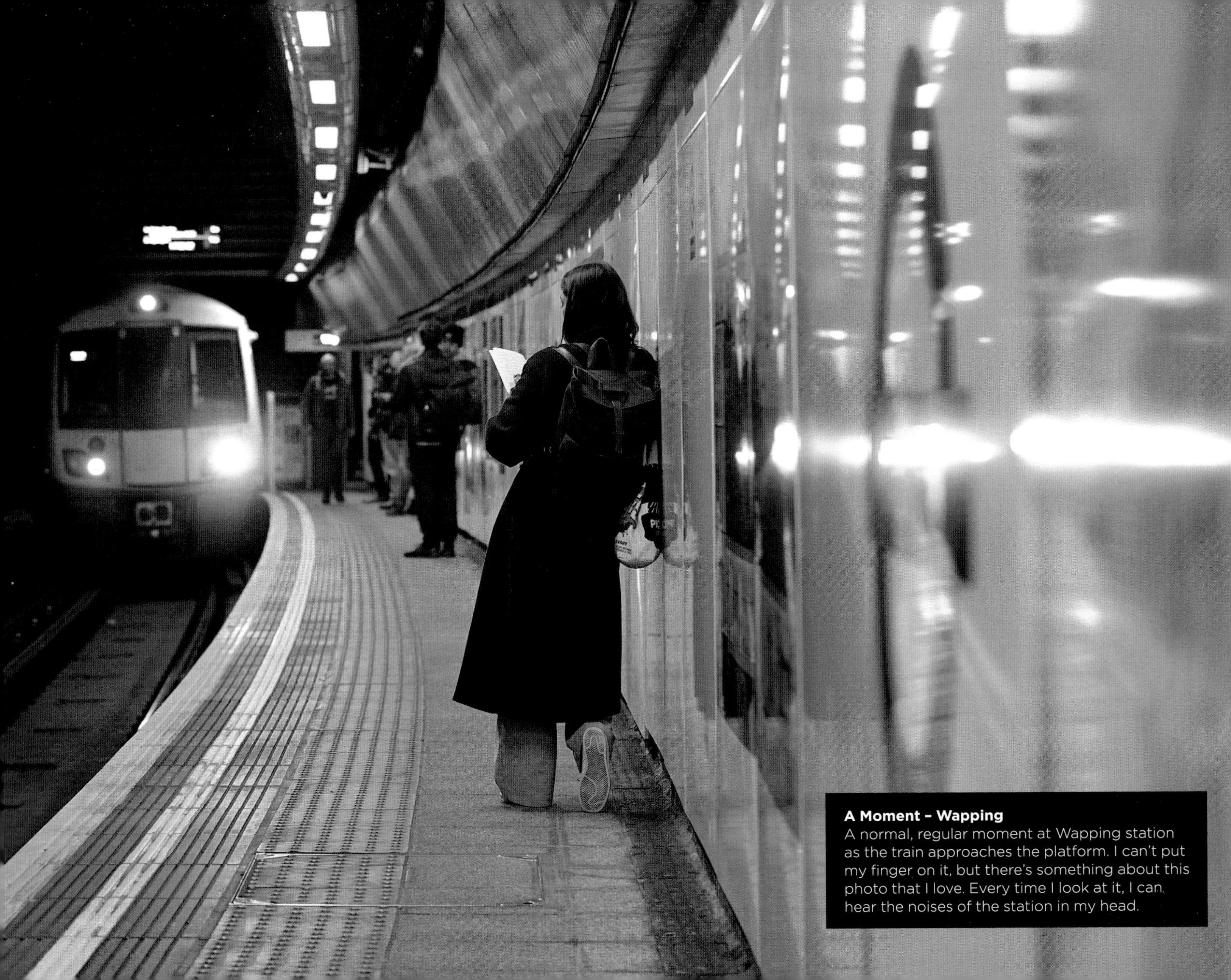

A Moment – Wapping
A normal, regular moment at Wapping station as the train approaches the platform. I can't put my finger on it, but there's something about this photo that I love. Every time I look at it, I can hear the noises of the station in my head.

ABOUT THE TUBE MAPPER PROJECT

There are currently 272 Underground, 113 Overground, 41 Elizabeth line and 45 Docklands Light Railway stations, as well as 39 tram stops. The Tube Mapper project aims to capture visual moments at each and every one.

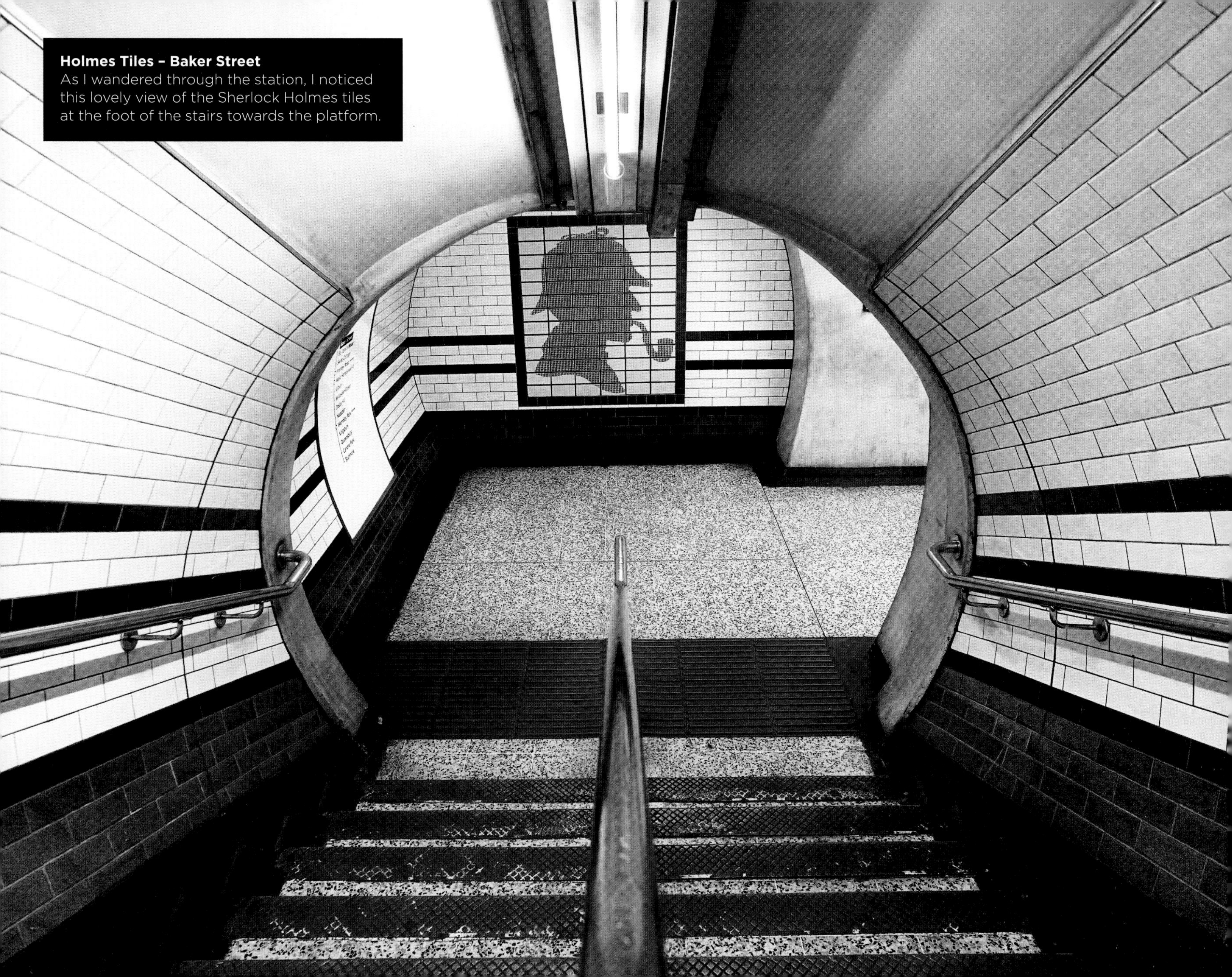

Holmes Tiles – Baker Street
As I wandered through the station, I noticed this lovely view of the Sherlock Holmes tiles at the foot of the stairs towards the platform.

FOREWORD

There is an inherent contrast in much about the London Underground: a dynamic tension between its past and present, between its heritage and its contemporary requirements. I had the privilege of playing a small role in helping manage the perception that they are competing requirements. Yes, first and foremost the Underground is an operational railway, a vital part of modern London life, and modernisation is needed – but not necessarily at the cost of its past. By necessity, as you would expect, the older the system gets, the more demanding that contrast becomes. It often goes on behind the scenes, such as fitting new twenty-first-century signalling through a Baker Street station that has been in use since the first Underground train steamed through in 1863. Sometimes these tensions, and the often creative compromises that working together can produce to respond to them, play out in public spaces. Sometimes the answer is to make the new an obvious contrast to the old; sometimes it is the reverse, making the new seem more comfortable with the old.

The fascinating thing about these changes and contrasts are that they are all part of the rich tapestry of the Underground and help make the Tube the unique institution it is. The juxtaposition of old and new, of light and dark, of under or on or over the ground, are all components in what has become the considered identity of the system and therefore of the city, of the people who use it and of the staff who enable it, the latter providing the personal element, providing another contrast to this solidly built infrastructure.

Contrast is also often a vital thing and something we have to strive for, such as in assisting visually impaired passengers to safely use the system. Contrast is also part of the diversity of the world around us and so something to be celebrated, as seen in this book. The wonderful thing about Luke's photography is that he brings his personal glimpses of the many and changing contrasts we all often see, fleetingly, on the Underground, and in doing so he helps capture something of the essence of the network, the city and its people. He often uses his creative skill and vision to bring even small contrasts more readily to our eyes. I hope you enjoy as much as I do Luke's take on what was my world for so many years.

Mike Ashworth,
Design & Heritage Manager at London Underground (LTM/LU 1992-2018)

INTRODUCTION

The journey, not the destination, matters ...

T.S. Elliot

The Tube Mapper project is my unexpected journey into the captivating world of the London Underground network. This ongoing personal creative assignment began as a response to becoming a new parent. I was worried there would be no more spare moments to continue my artistic photography, which at the time involved photographs of cityscapes at sunrise and sunset. So, I instead focused my creativity on a new path: my overlooked travels to and from work. This new focus suited my new schedule better, and I slowly built up a collection of images that documented my encounters during my seemingly mundane commute on the Tube.

   The project grew in popularity over time, garnering attention from social media and news outlets. The increased interest in my photography led me to organically expand the documentation to include every station and stop on the Tube map. I have recently added the tram network to my adventures, and I'm also considering including all of the bus stations in London. My love and enjoyment of the journey have shown me that, the closer I get to reaching my destination, the more my eyes open up to the hidden delights there are to discover along the way.

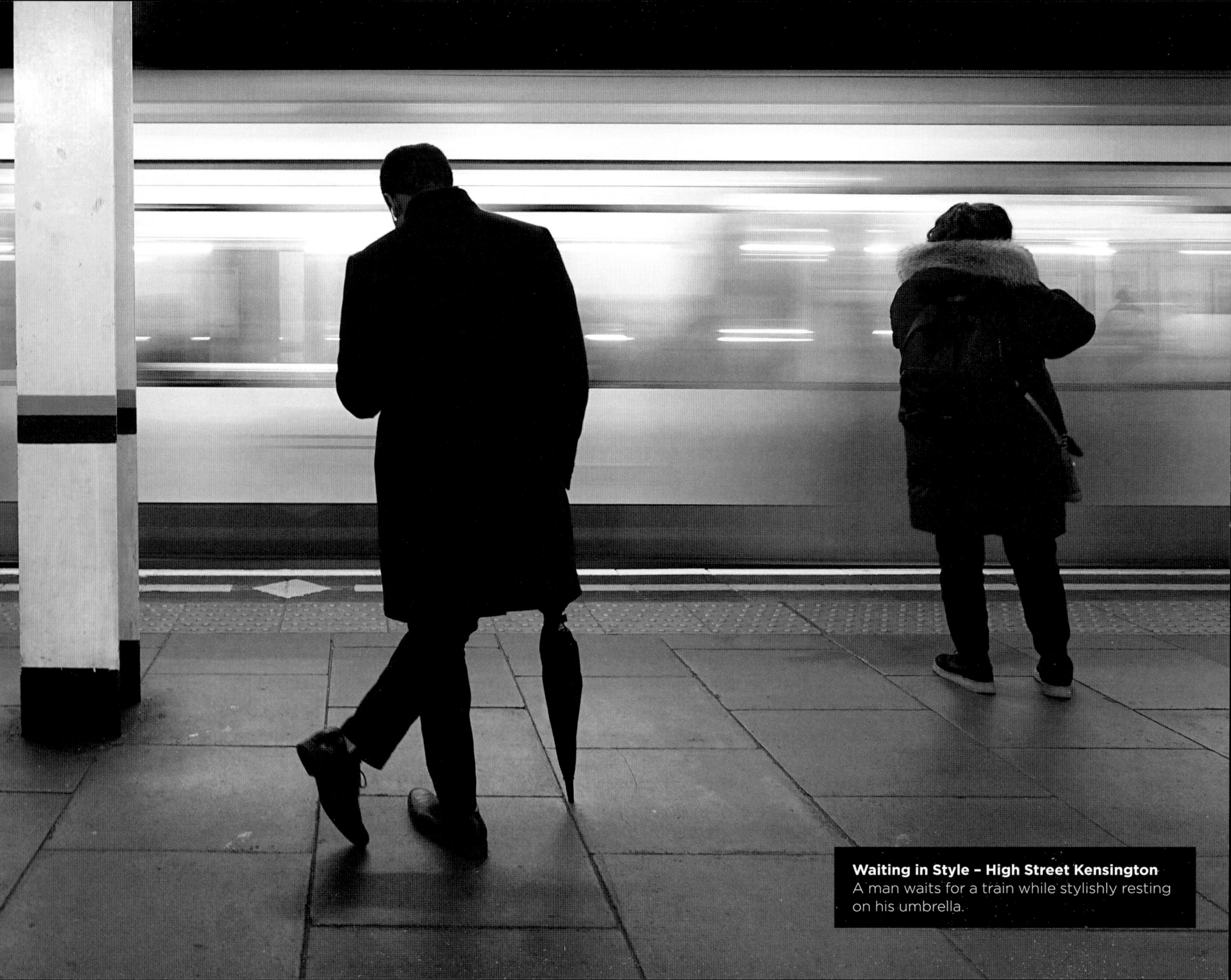

Waiting in Style – High Street Kensington
A man waits for a train while stylishly resting on his umbrella.

CONTRAST ON THE UNDERGROUND

The London Tube network is the world's first underground railway, which opened in 1863 as a single underground line from Paddington to Farringdon Street. Over 160 years it has evolved into a complex transport infrastructure, reaching more than 4 million daily users in 2023. The transformation in complexity highlights its impressive ability to adjust to the ever-changing needs of the people it serves, creating a beautiful visual juxtaposition of modernisation and ingenuity in some of the oldest underground transport structures in the world.

The Underground is a world of contrasts. It is a giant museum that celebrates the old and new and spans the city of London. As you navigate the rabbit warren of historic and modern architecture, it's impossible to not stop and take notice of the contrasting elements that many overlook during their daily travels.

For example, there are over 450 escalators on the Underground, and every week these 'moving walkways' travel a greater distance than the circumference of the Earth. It is now unimaginable to think of a station without them, as escalators have become a core function of how these buildings function. Yet, when underground stations were first constructed, these electronic staircases did not exist. Their inclusion can provide a fascinating tonal shift to the overall style and architecture of the station where they reside.

The reactive nature of the Underground and connecting services is what makes the network so special. Growing populations demand greater capacity within existing stations and require increasingly more frequent functioning links to stations outside the city centre. On 24 May 2022, the Elizabeth line opened: a 73-mile (118km) railway line in south-east England that travels through London. It has reduced congestion in Central London by increasing rail capacity by 10 per cent and, on average, there are over 600,000 journeys made on it every weekday, reaching 4.5 million users in the last week of September 2023.

Visually, the Elizabeth line stations are dissimilar from their Tube counterparts. Their spaces are vast in scale, with sleek, modern and elegant architecture. Even the trains are different, being over 200 metres long, which is almost twice the average length of an Underground train. What is impressive is how TfL integrated the Elizabeth line with their other services, enabling travellers to seamlessly navigate through these contrasting spaces. Despite their structural differences, all feel like part of the same network. This approach of maintaining and expanding its services while keeping true to its brand identity is one of the things that I most admire about TfL. They have a 'Design Idiom' that outlines their ethos towards all aspects of their construction:

> The Idiom has been put together to recognise, conserve and sympathetically nurture the design heritage that already exists and to inspire great design in all new projects, regardless of scale.
>
> TfL Design Idiom by Studio Egret West, 2016, p. 4

This transport network's promise to preserve the soul and heritage of one of the UK's most iconic brands is one of the features that makes it so unique. The effort that goes into protecting the design standard of each station is appreciated, making it a pleasure to photograph each location. This book documents my exploration of the photogenic contrasting elements that I encounter as I visit every station and stop on the London Tube Map.

1

UNDERGROUND/OVERGROUND

Older readers will surely remember The Wombles, but did you know the surprising fact that only around 45 per cent of the Tube network is underground, with the rest exposed to the elements? This percentage dramatically decreases when considering that my photography project captures all stations and stops, including trams, the Docklands Light Railway (DLR), and the aptly named Overground network; this means most places on the map are above ground.

So here is a collection of photos enjoying the differences between underground and overground locations.

Photo Hints

Underground: Shots looking into tunnels are taken safely with a telephoto lens while standing behind the yellow line. Looking for curved platforms helps, as you can sometimes view straight into a tunnel.

Overground: To discover exciting vantage points, wander outside stations and look for nearby bridges and elevated structures. You can also discover compelling viewpoints from inside trains, so keep an eye out on your journey.

Coming and Going – Westminster
I enjoy watching trains travel through dark tunnels and the art that they create. The paint palette is mainly two colours: the bright white/blue from the train's headlights and the red from the back.

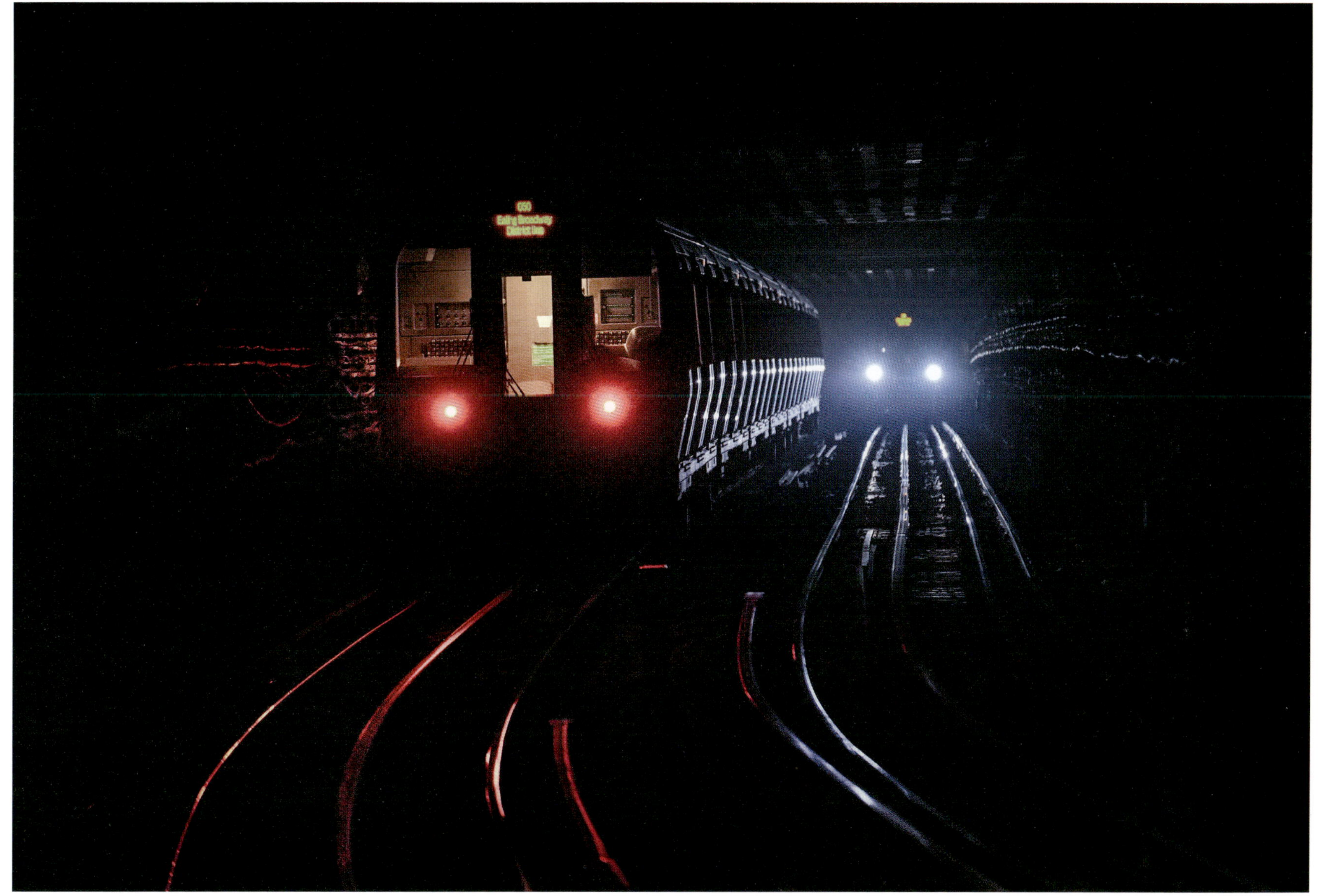
050
Ealing Broadway
District line

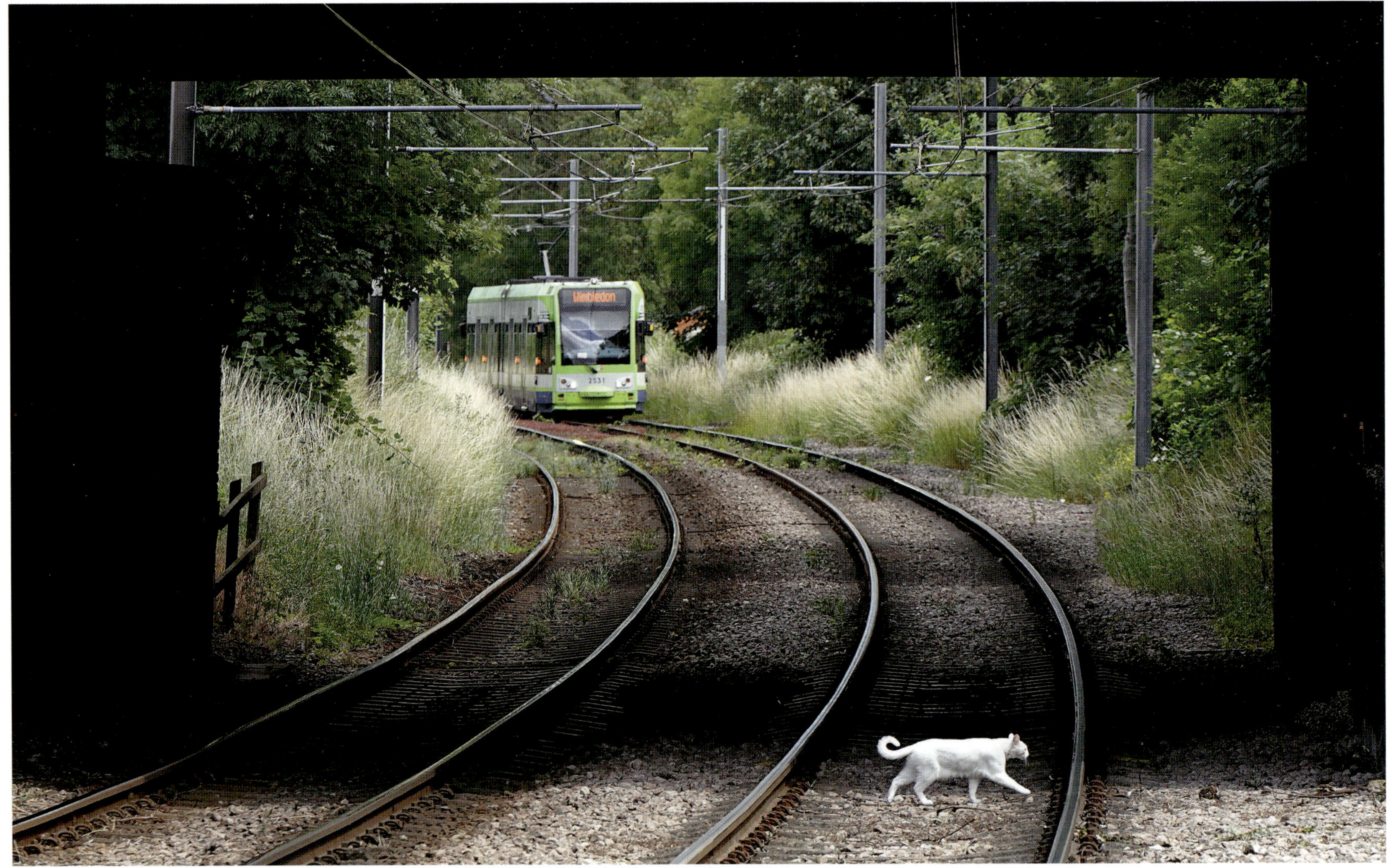

Catwalk – Woodside
Sometimes you get lucky with photography. I was already in focus, waiting for a Croydon tram to arrive at Woodside, when suddenly a brilliant-white cat walked across the track. It was a lovely moment.

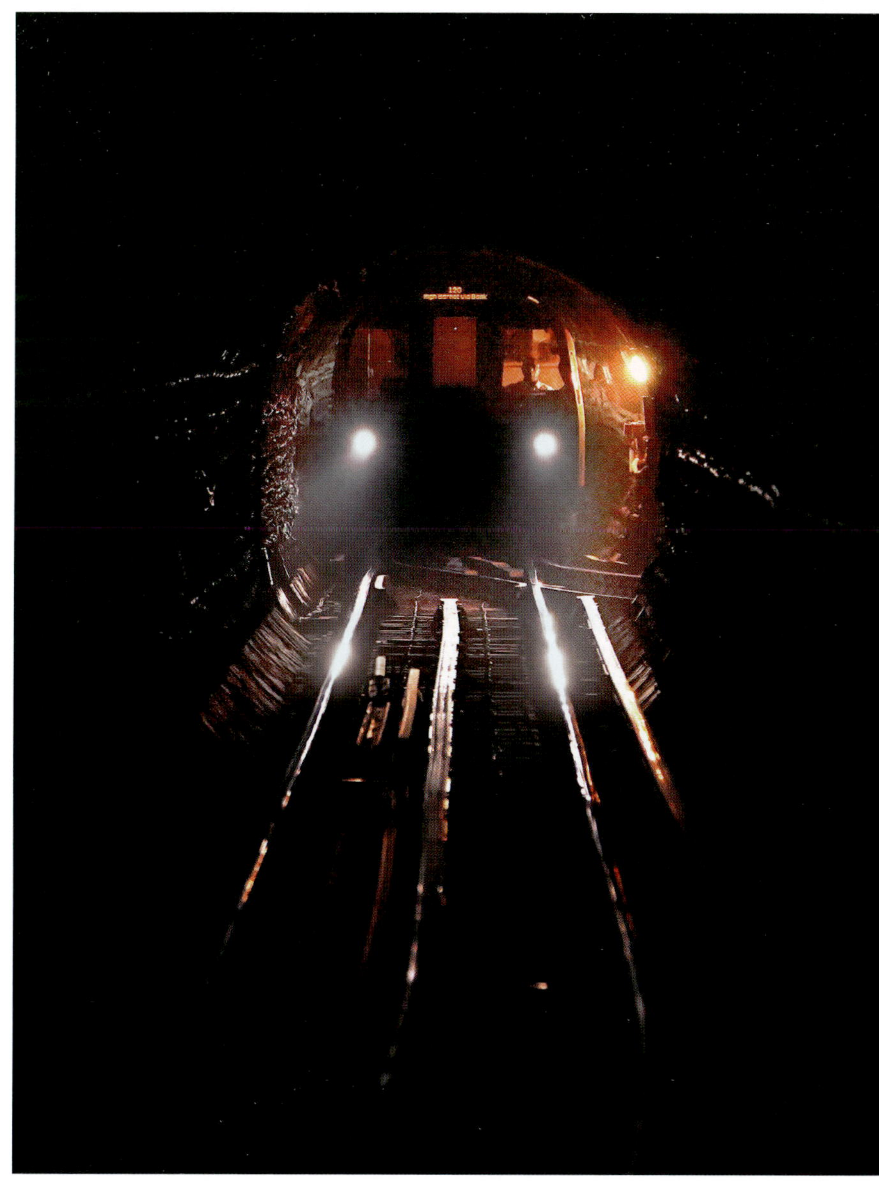

Ghost Train – Kennington

This spooky scene is at Kennington station, infamous for the haunted Kennington Loop, where the train turns around and passengers aren't allowed on. The track has had a ghost since an incident in the 1950s. Apparently, even some drivers don't like this part of the Underground.

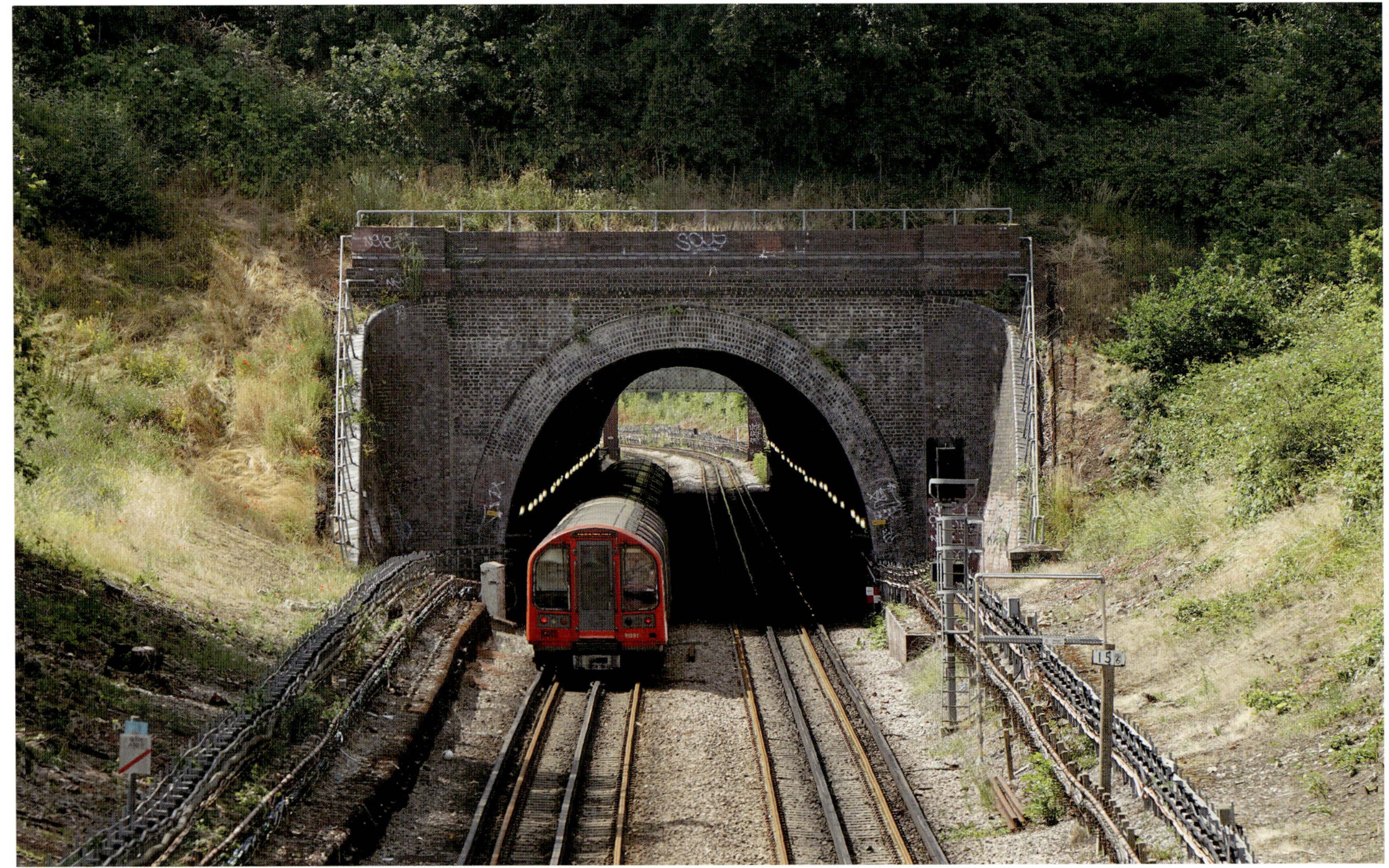

Through the Hill – Grange Hill
If you are as young as me (or slightly older), you may link this station's name to an old BBC television series about a school.
Opposite the entrance is a lovely vantage point looking down on the track from where you can watch trains travel through the tunnel in the hill.

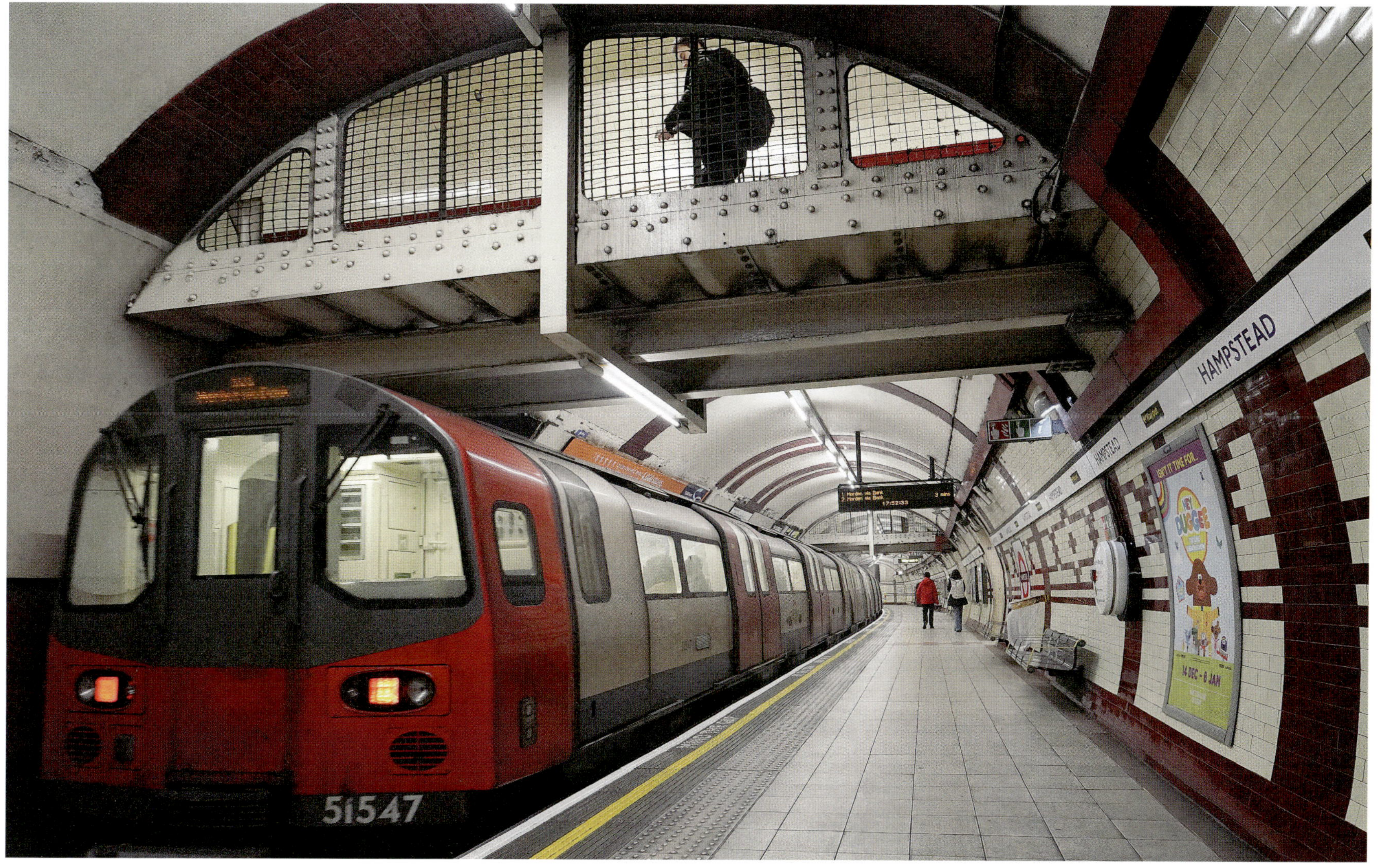

Walk Over – Hampstead
I love that you see can see people walking over the trains at Hampstead station.

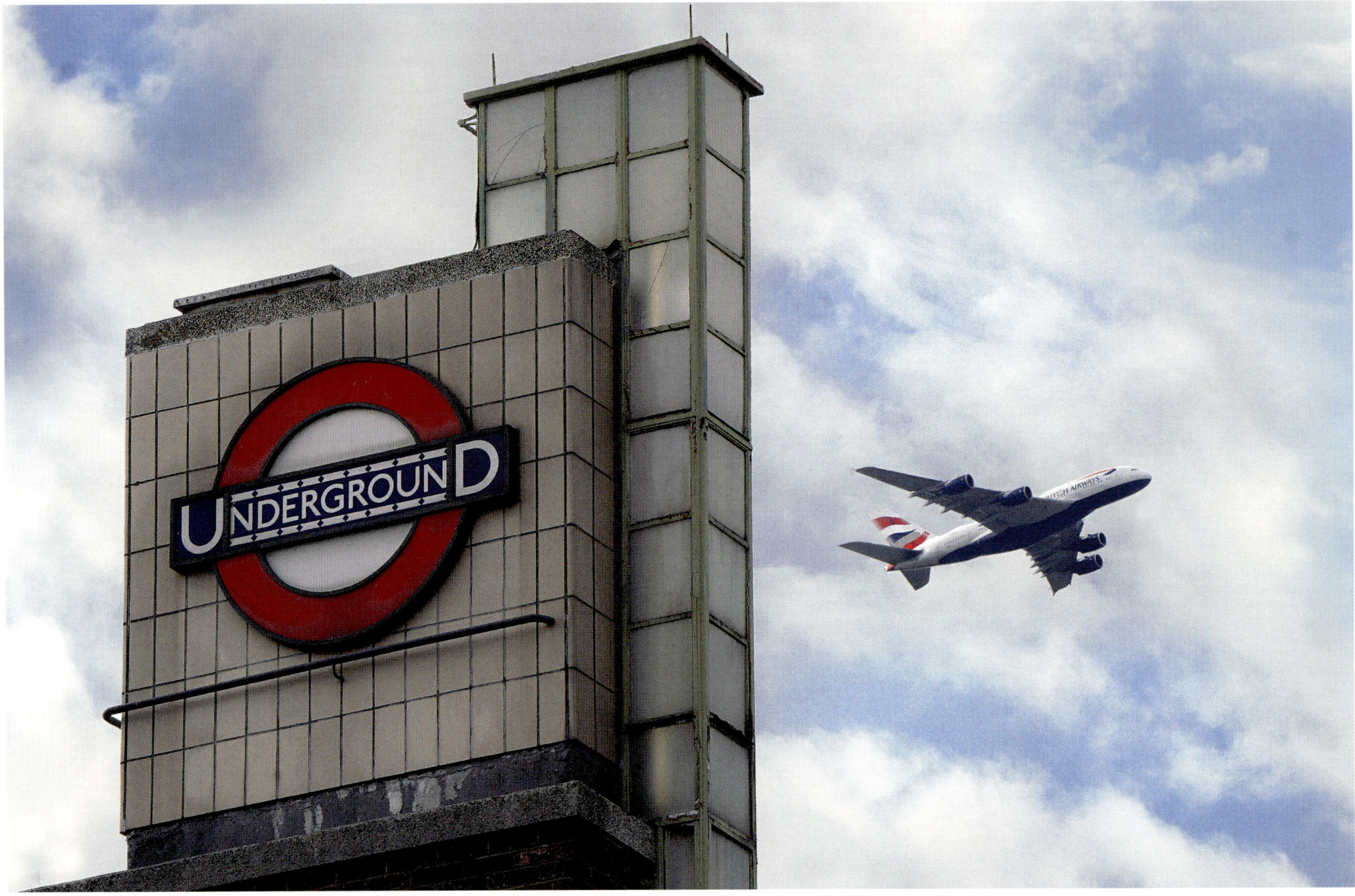

Flight to Boston – Boston Manor
I was travelling back from Heathrow and noticed that Boston Manor station was in the flight path. So I waited patiently, focusing on the London Underground tower, to capture a photo of a British Airways plane flying by.

Limitless

Stars fly past our Earthly home
Soaring out into the universe
No limits, no bounds
We are a humble place

Technological wonders surround us
Above ground
Under ground
We place our essence, our future there
We are free
We are limitless

Tracy Meyerhofer

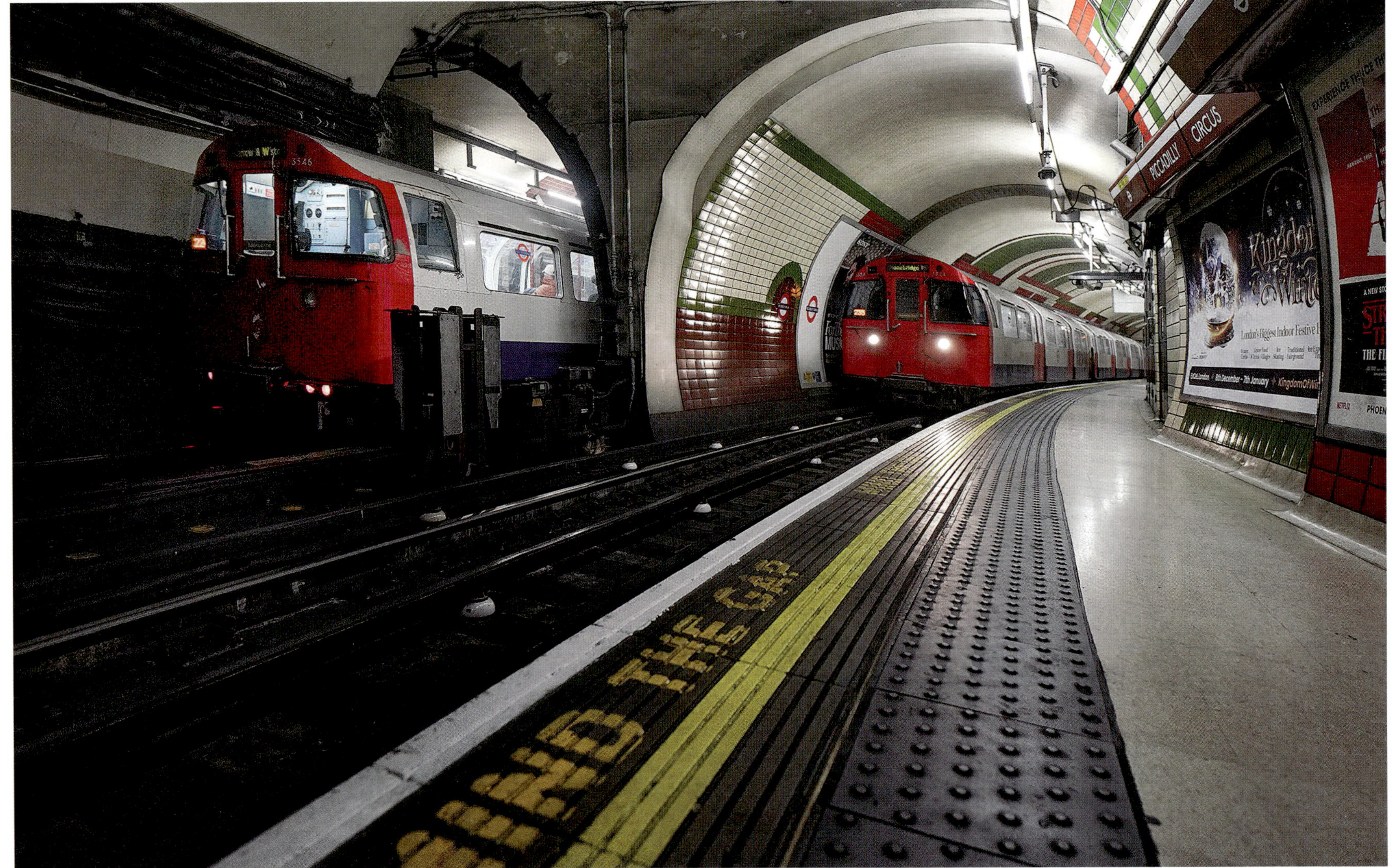

Twin Tunnels – Piccadilly Circus
A unique view from the end of the Bakerloo line platform at Piccadilly Circus station. It feels like you have an extra-special viewpoint from where you can see trains entering and departing from their respective platform tunnels.

The Late

There's a gap in my mind
a dark space, near black.
A space once occupied
by something light,
something human, I think,
sheltering in history.
I'll call it 'you'.
There's a gap in my mind
once full of many yous
all happy and bright,
a smile, sunny and sepia,
now deep underground,
once soaring and full of laughter.
There's a gap in my mind
leaving part of me on a platform
where doors closed far behind,
a memory down there at midnight.
There's a gap in my mind
a message warns me
and I really really do mind.

Danny Herbert

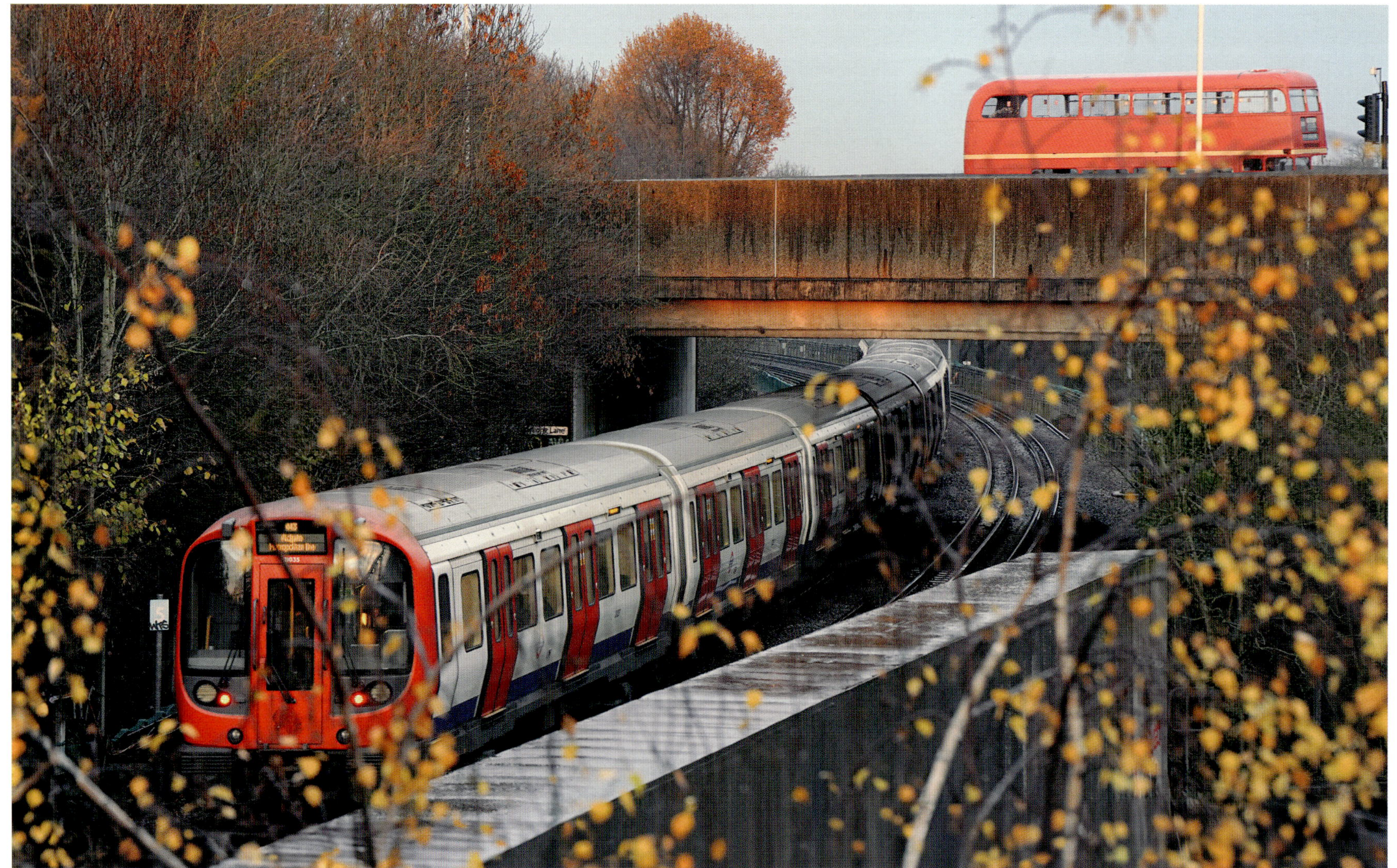

Above, Below – Hillingdon
This is a photo of the beautiful autumnal colours from a bridge at Hillingdon station.
A glorious off-frame sunset and a passing Routemaster bus on the bridge above the train further intensify the colours.

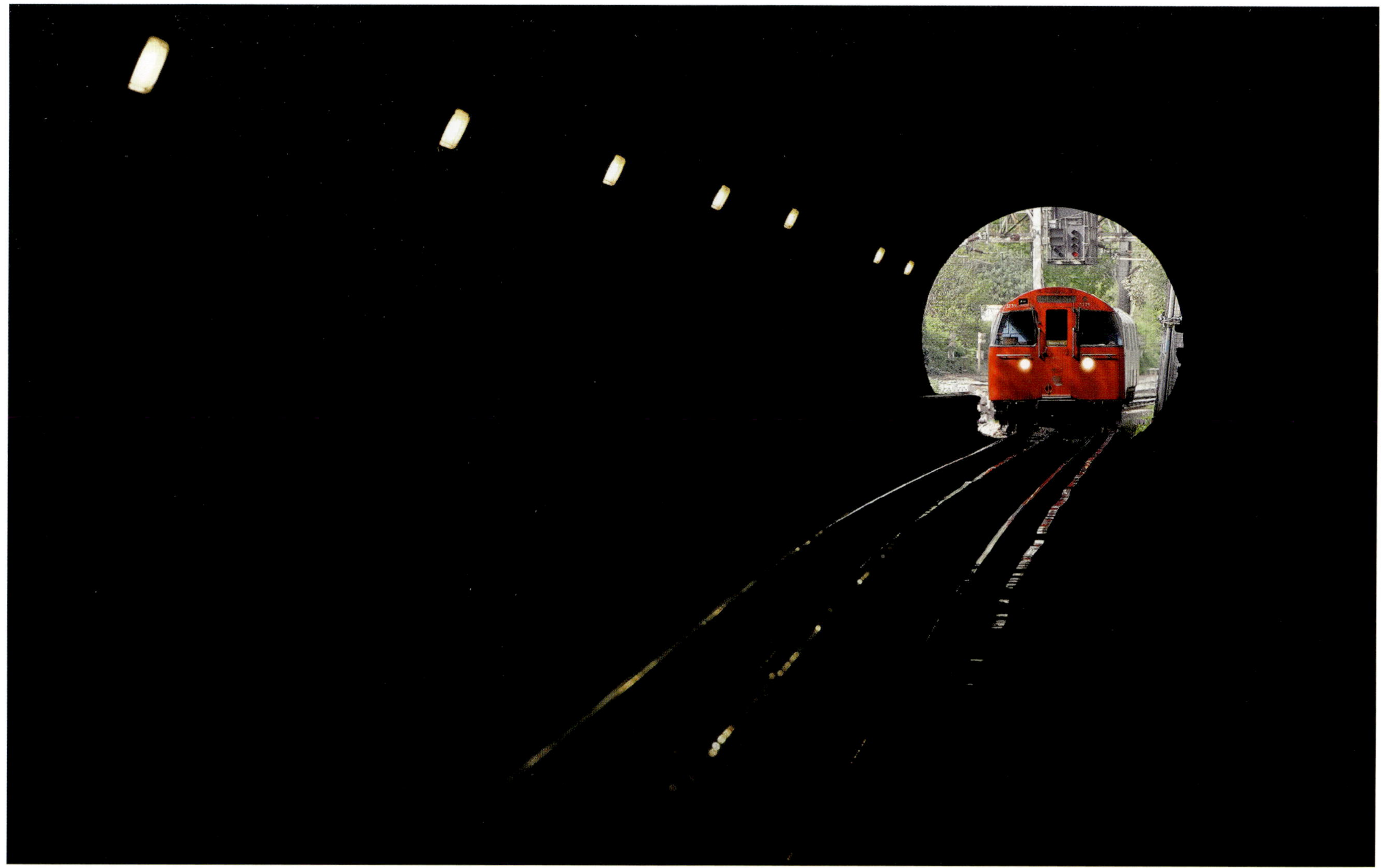

Thinking about Trains – Kensal Green
From behind the yellow line on the platform at Kensal Green, you can see trains arriving through a long tunnel as they approach the station.
The lights in the dark tunnel look like cartoonish thought bubbles for the Bakerloo line train.

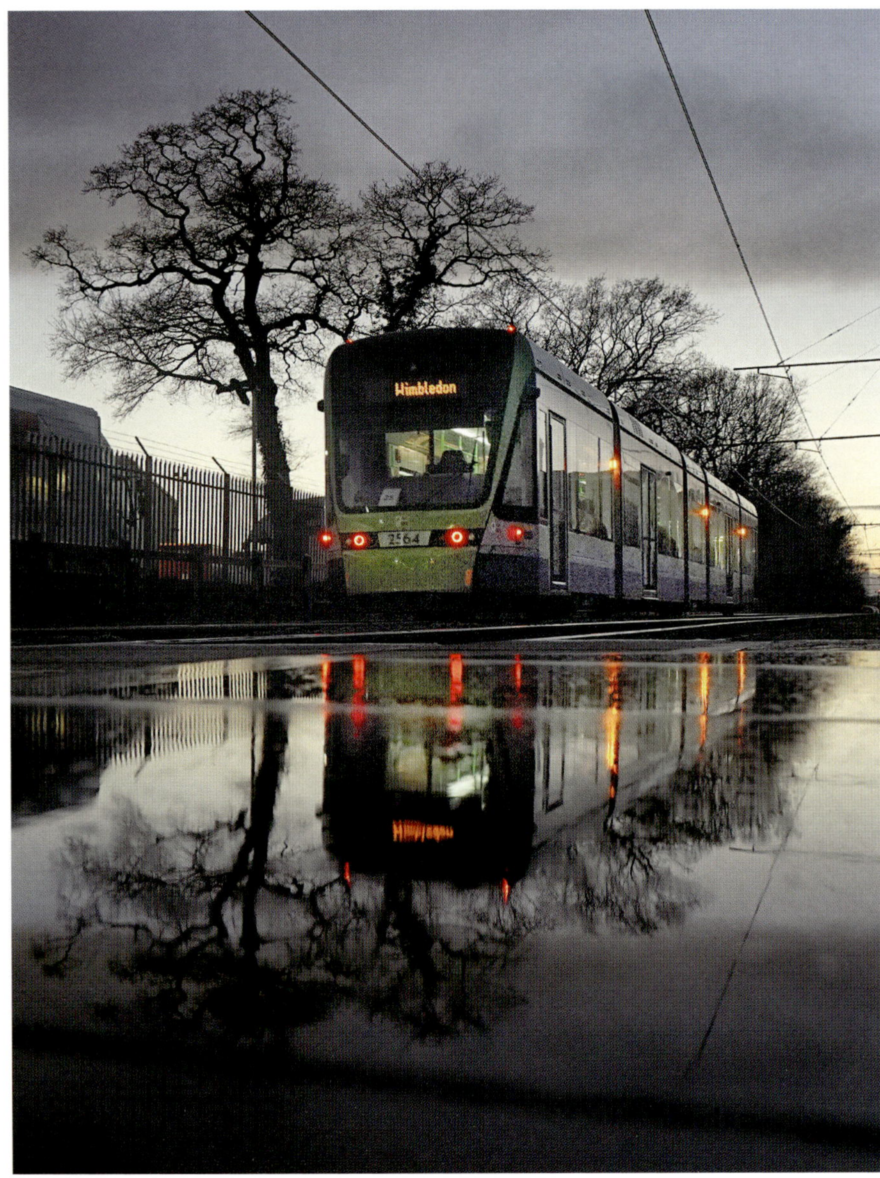

Trams at Sundown – Beddington Lane
The sun had set on a rainy evening at Beddington Lane tram stop.
The nightlights of the departing tram, reflected in the puddle
below, look magical.

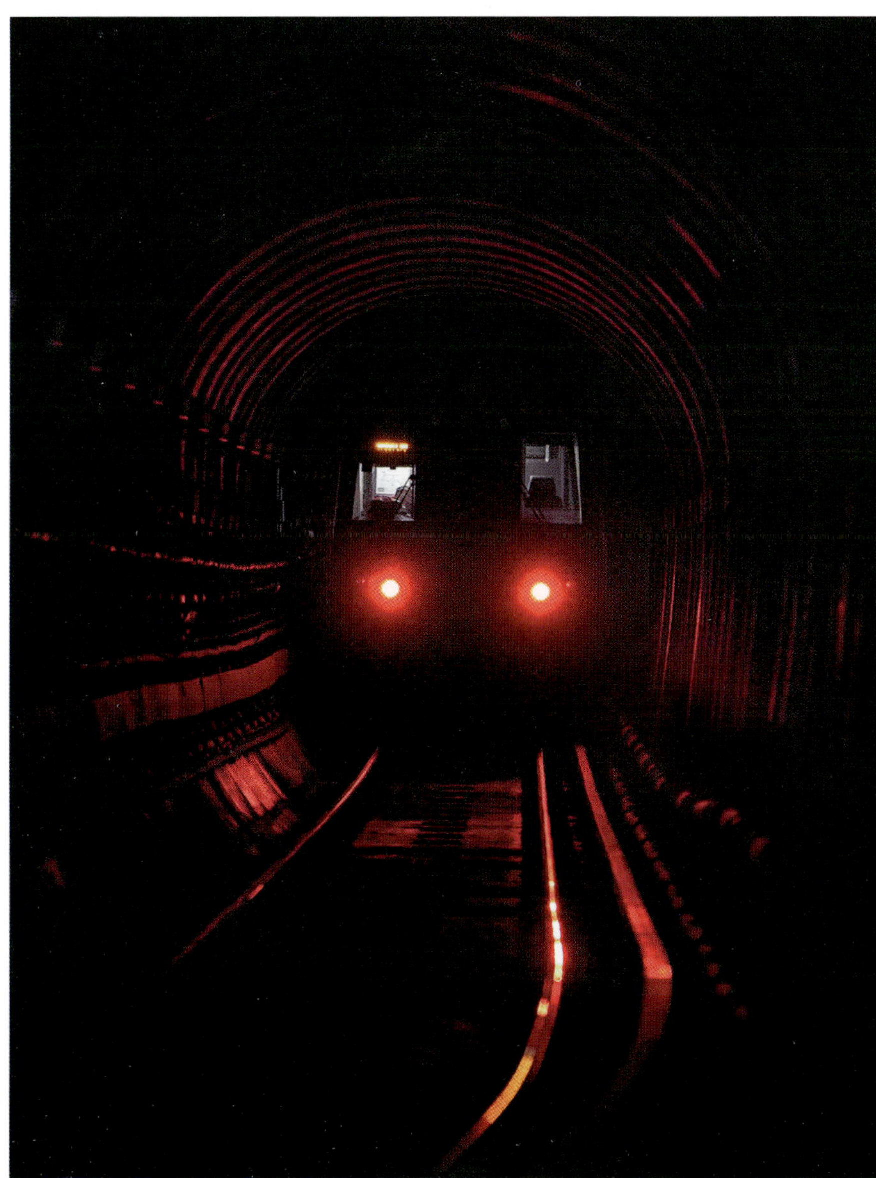

Seeing Red – Wapping
A departing train lighting up the historic Thames Tunnel that Marc Isambard Brunel designed. The first underwater tunnel to be built anywhere in the world, it opened to pedestrians in 1843 and would change into a railway tunnel in the 1860s.

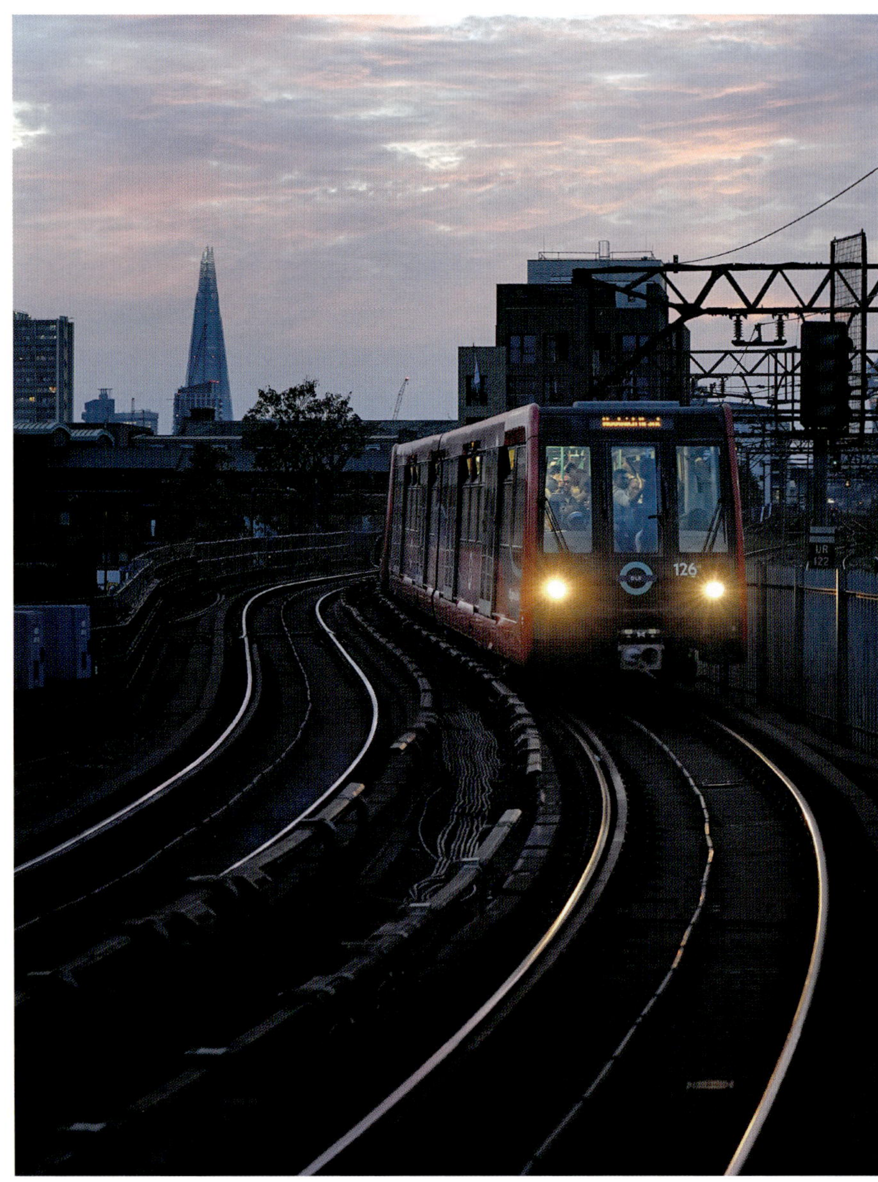

Pink Sky – Limehouse DLR
A DLR train arrives at Limehouse station, with a lovely backdrop
encompassing the Shard. The sun was setting, and I was
fortunate to have a pink sky.

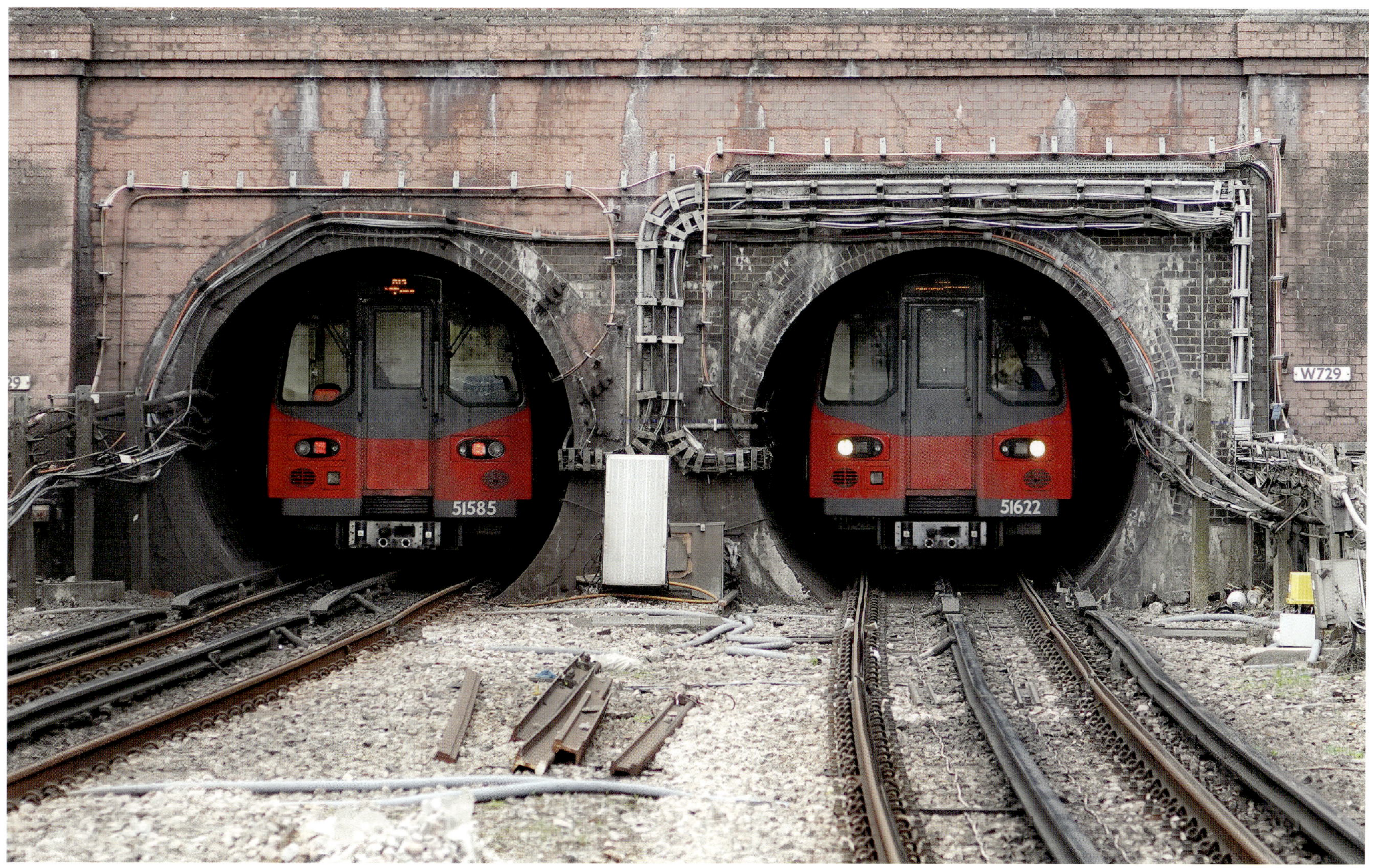

The Perfect Fit – Hendon Central
This above-ground Northern line station offers a fantastic view of a pair of Underground tunnels from the end of the platform,
looking northbound towards Colindale. This is a visual articulation of why we call it the Tube!

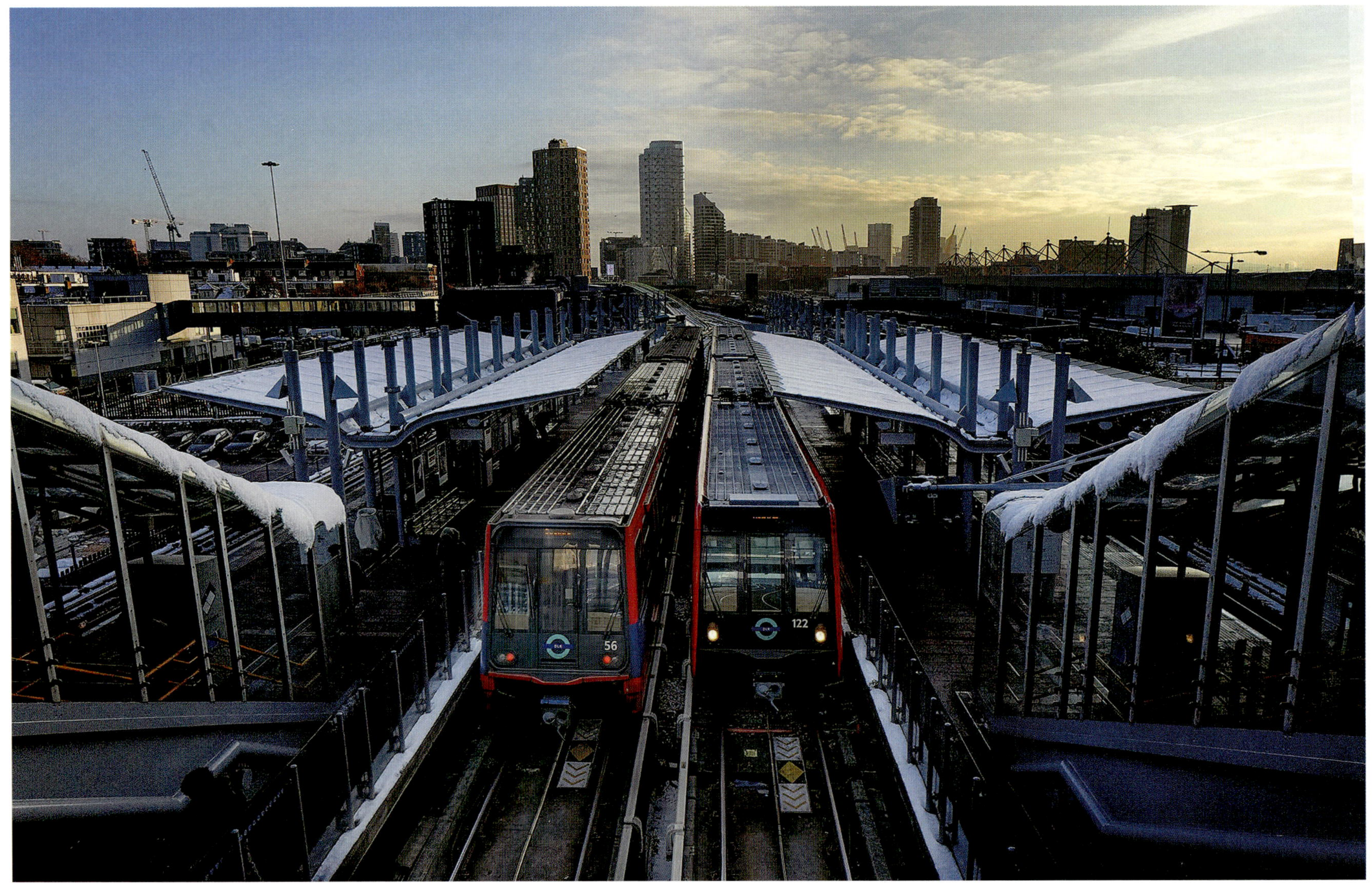

Popular Trains – Poplar
A great view of the City and two trains docked on a snowy morning, taken on the footbridge that leads to the entrance of the Poplar DLR station.

Return Ticket

Use the Tube going one way
From Victoria to Bank, say

Then follow the same track
Above ground on the way back

Victor Keegan

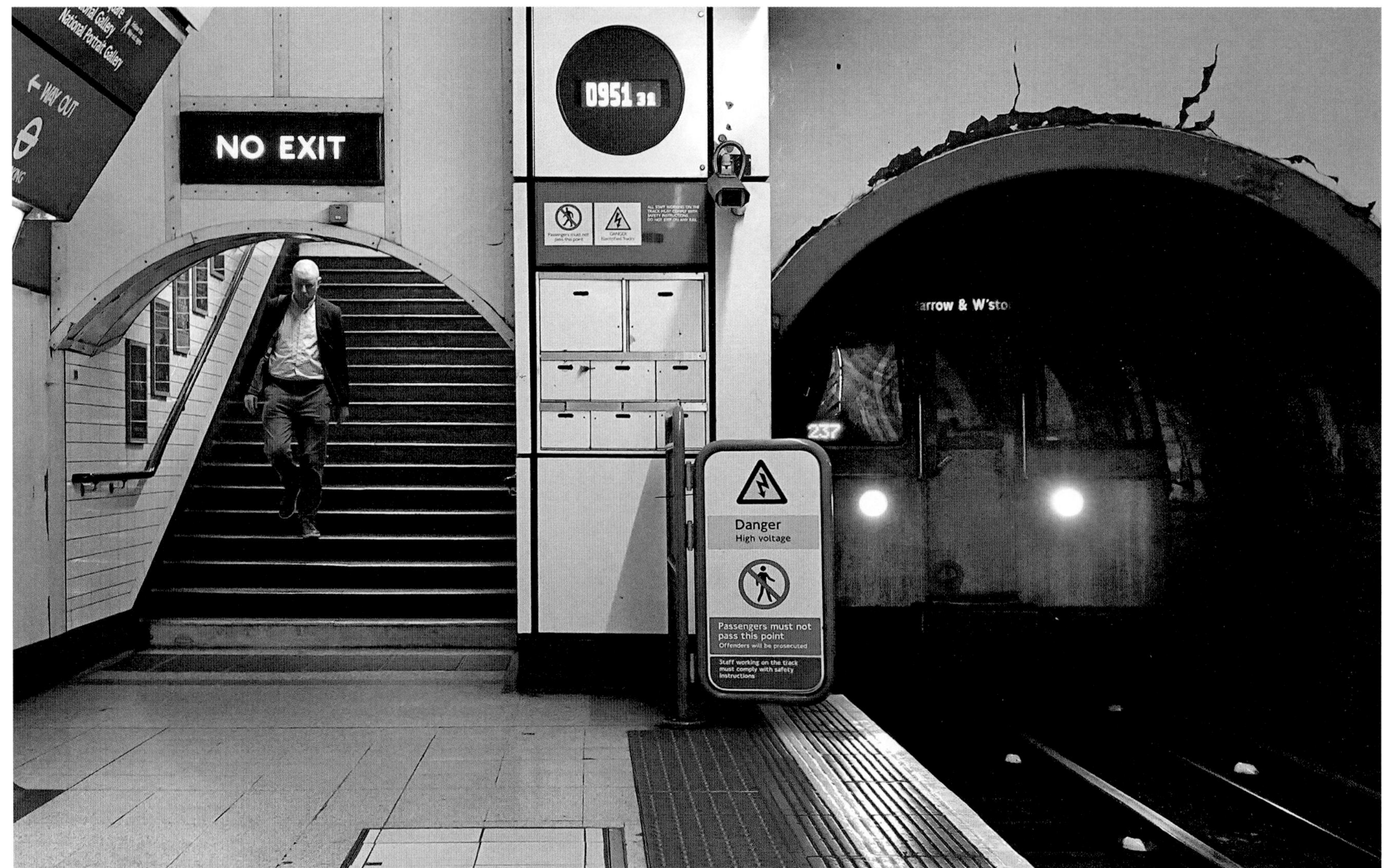

Exiting Tunnels – Charing Cross
A commuter and a train coming out of their respective tunnels.

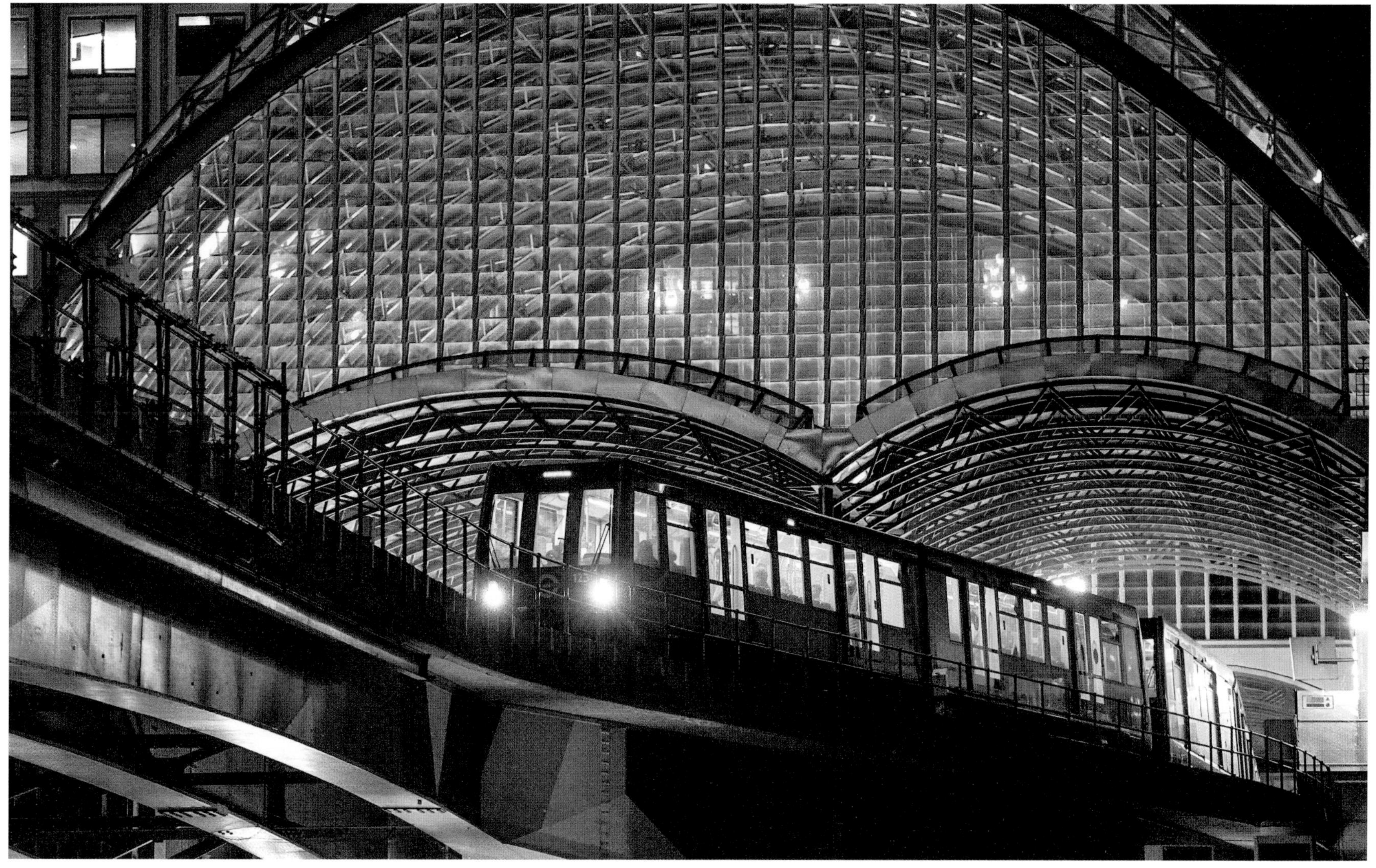

Night Lights – Canary Wharf
A DLR train heads towards Heron Quays station at night. I'm a big fan of the Canary Wharf DLR building in the background and I love the lights shining through its curved canopy structure.

Out of the Tunnel – Island Gardens DLR
A DLR train enters Island Gardens station through a dark tunnel. I love how the lights
paint the track in the darkness and especially enjoy the arching light framing the tunnel.

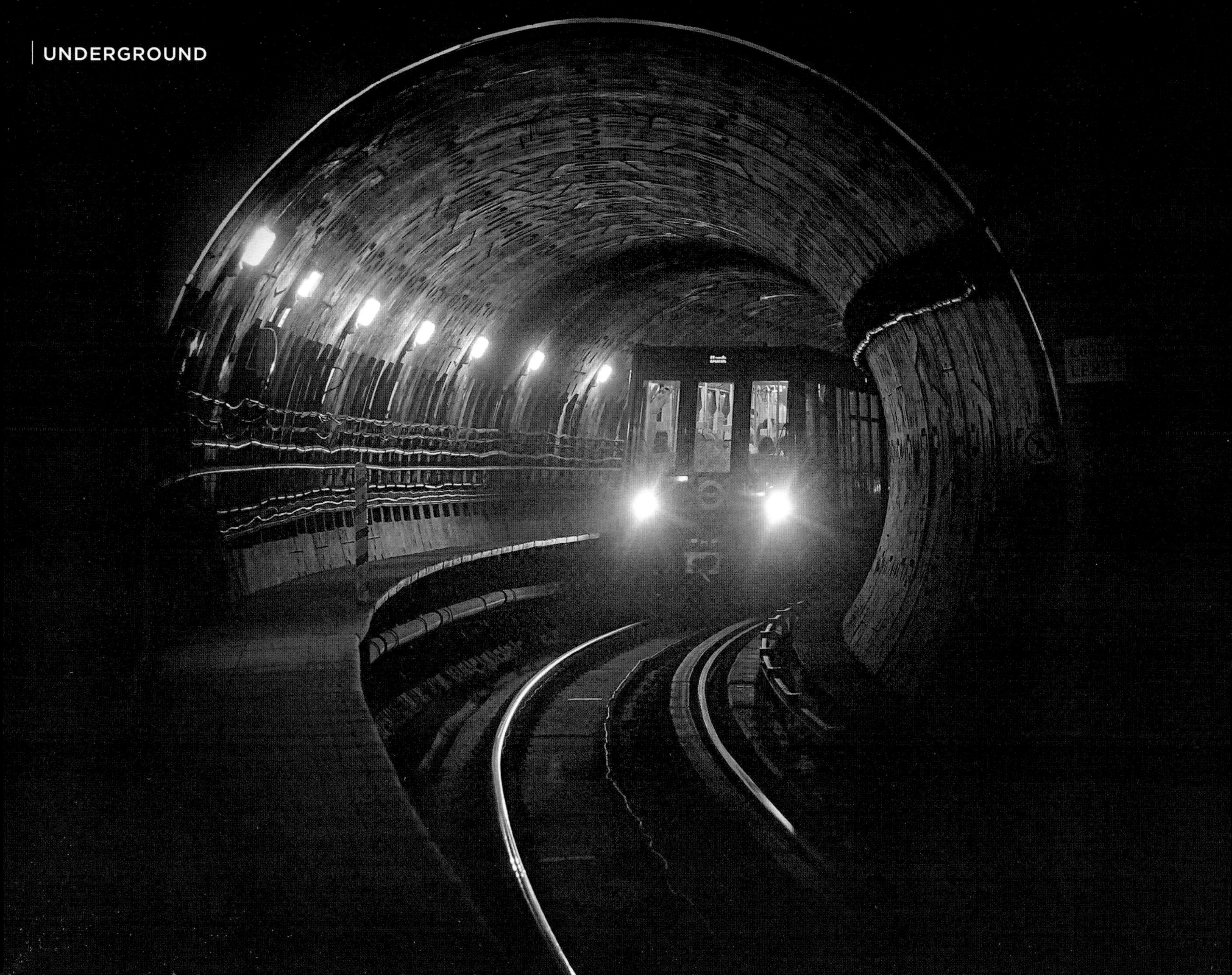
UNDERGROUND

A Train Approaches – Penge West
A London Overground train arriving at Penge West railway station.
I enjoy the dappled light and the contrast of the foliage with the metal bridge.

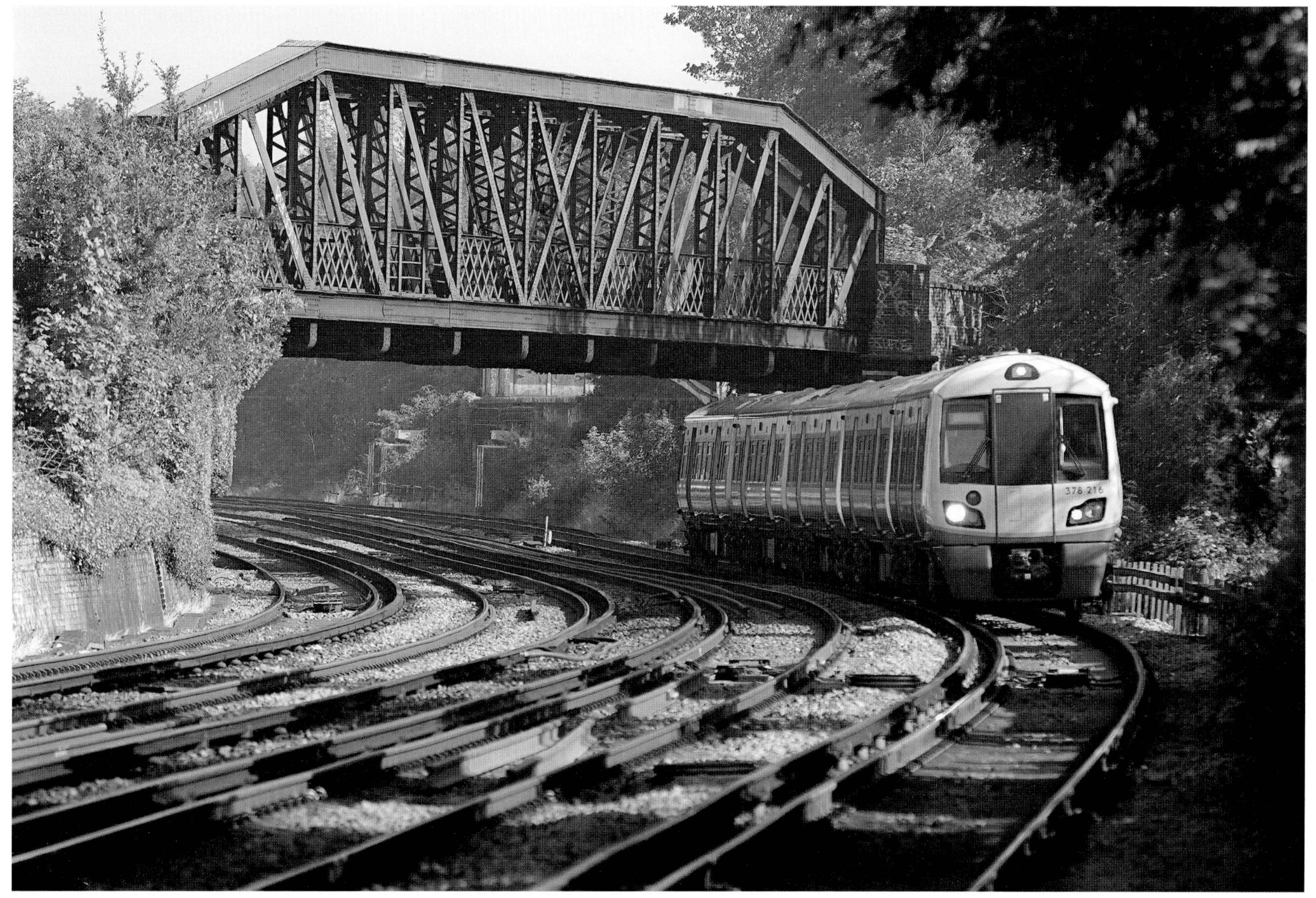

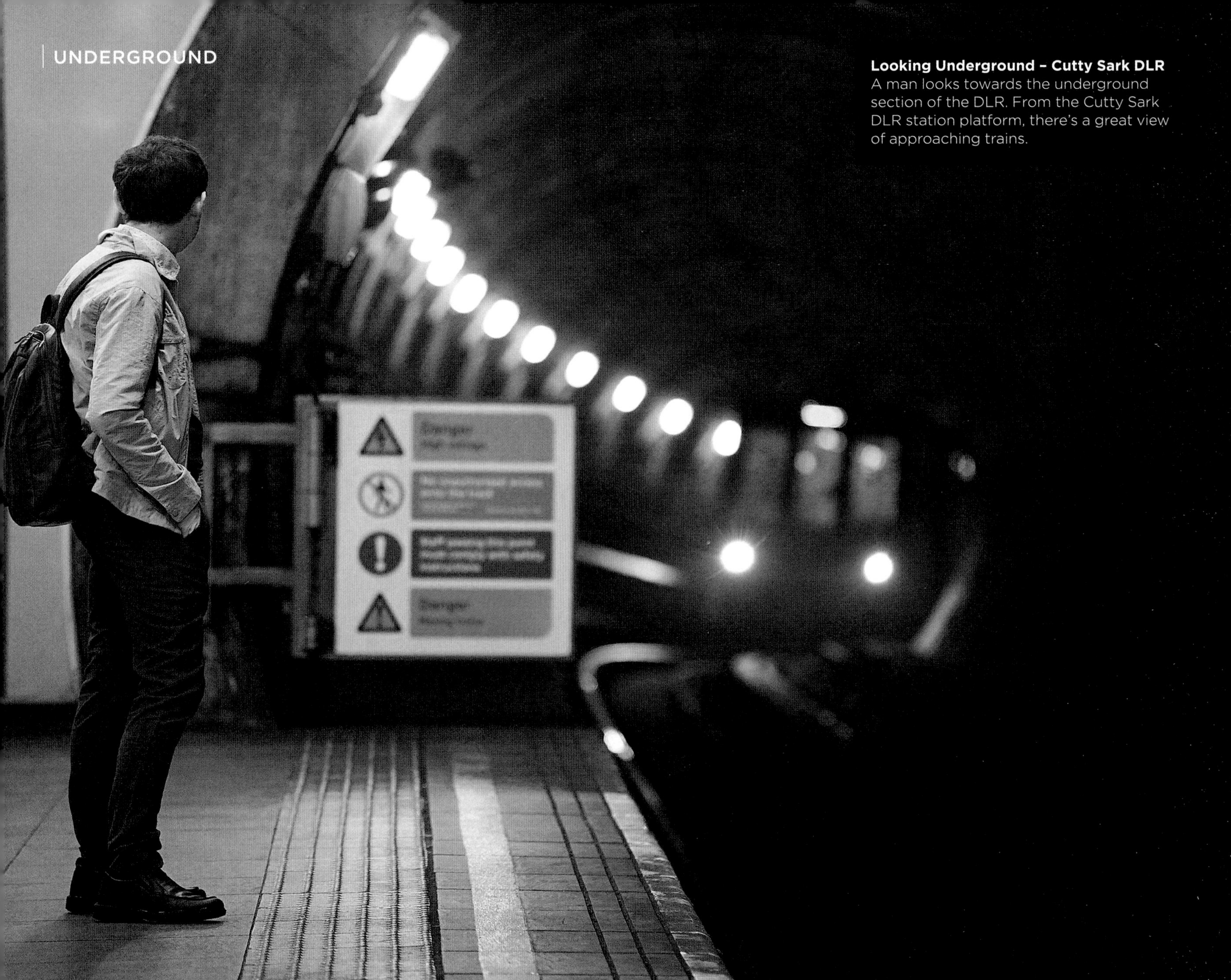

Looking Underground – Cutty Sark DLR
A man looks towards the underground section of the DLR. From the Cutty Sark DLR station platform, there's a great view of approaching trains.

Directions

Eyes meet on opposite platform,
in the tunnel a spark
briefly illuminates the dark,
a midnight thunderstorm.

A smile across the tracks
interrupted by trains
taking them in different lanes
with hearts of melting wax.

Danny Herbert

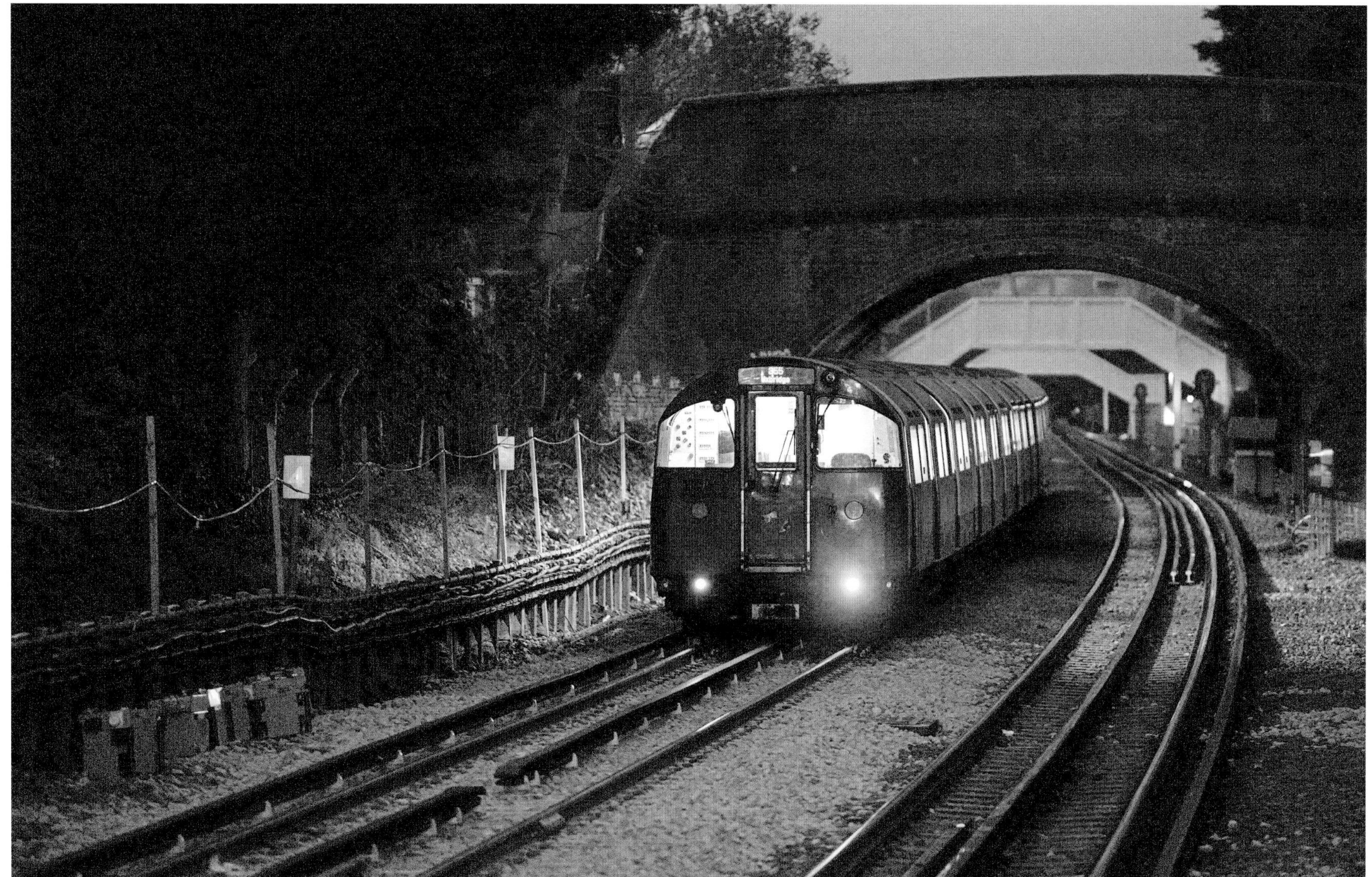

Dark Departure – Sudbury Town
A London Underground train leaves Sudbury Town station late in the evening.

2

FUTURISTIC/TIMELESS

The London Underground grants every user the ability to become a time traveller. Walking down a hallway surrounded by old signage and original features can magically transport you back to the early 1900s. We can taste a world that existed over 100 years ago and see hints of how our society functioned before today's technological advances. But once we ride a few stops on a train or walk to a different station area, we can suddenly feel like we are roaming a futuristic society with shiny walls, metallic buttons and modern architecture.

Photo Hints

Timeless: For truly timeless images, try to avoid obvious twenty-first-century technology. Hiding mobile phones and digital screens helps with the illusion.

Futuristic: There are many throughout the network, but if you want a quick dose of futuristic-looking locations, visit all stations on the Jubilee line between Westminster and North Greenwich.

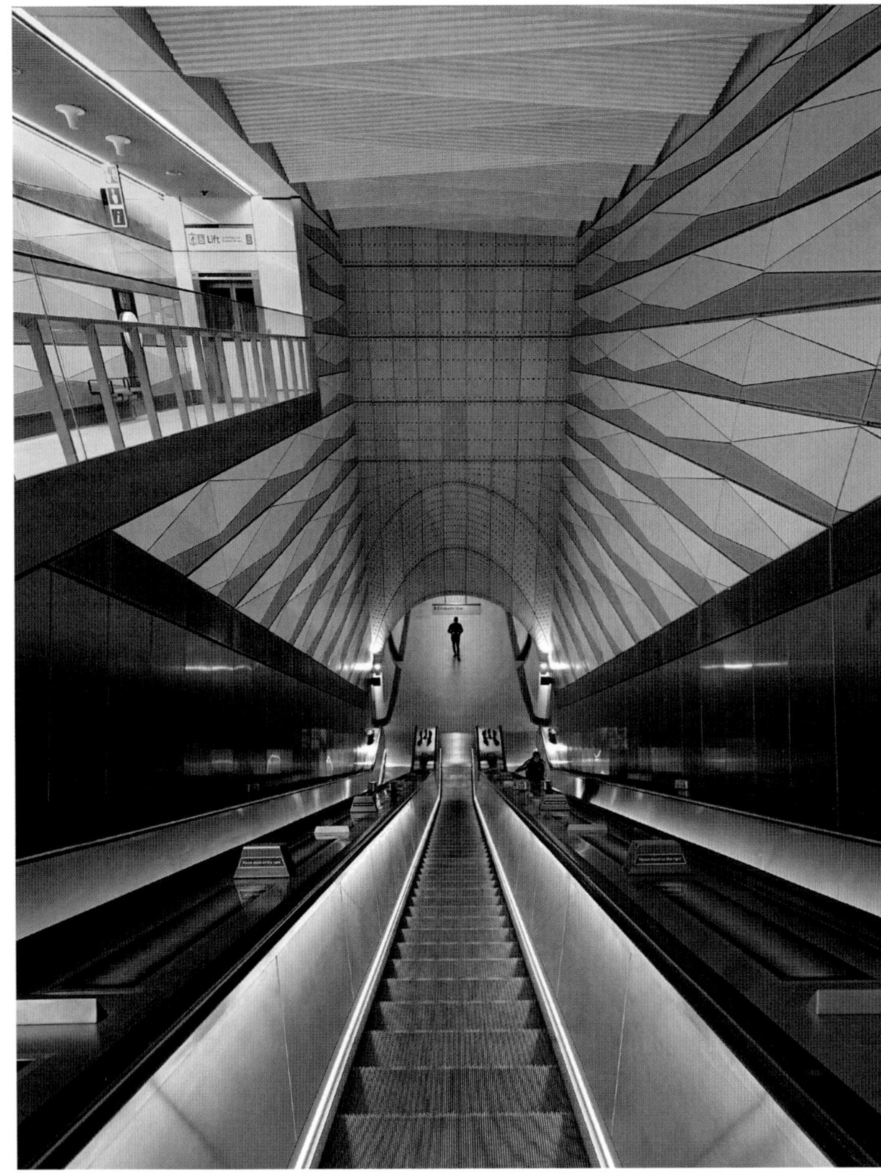

Elizabeth Lines – Liverpool Street
The dramatic architectural view of the escalators at this Elizabeth
line station; there are so many textures and patterns on display that
you immediately feel like you've stepped into the future.

Dreaming of tomorrow
Light bounces off the dark curve of the tunnel,
Before the train emerges, a time traveller,
Or a glimpse of travel in the future.
On the platform waiting passengers
Grapple with the modern day
Hoping the train carries the answers.

Danny Herbert

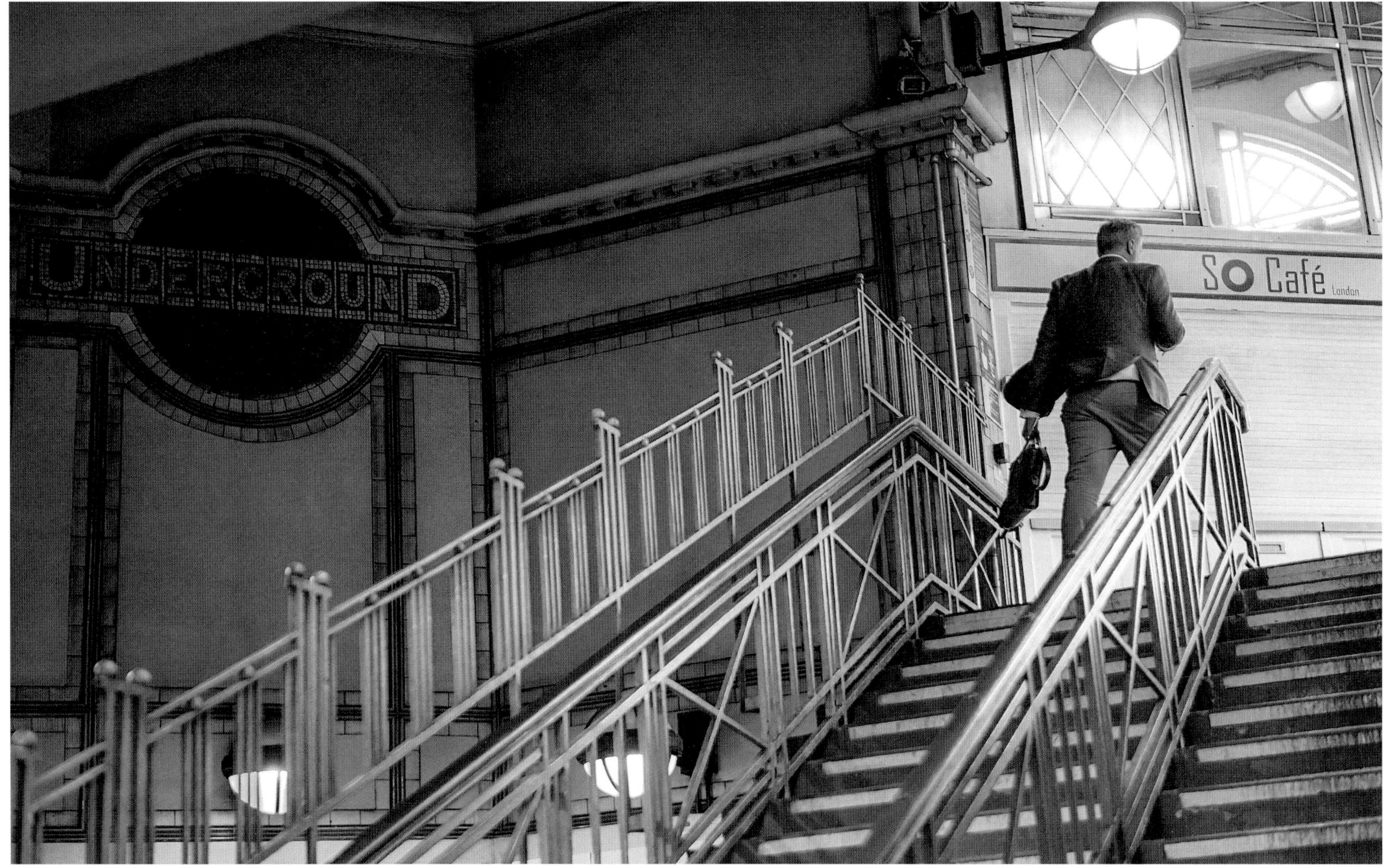

Leaving the Station – Maida Vale
This 1915 Grade II listed building won a National Railway Heritage award in 2009 for its modernisation of a historic station.
The stand-out features are the two large mosaic roundel logos at the station's entrance.

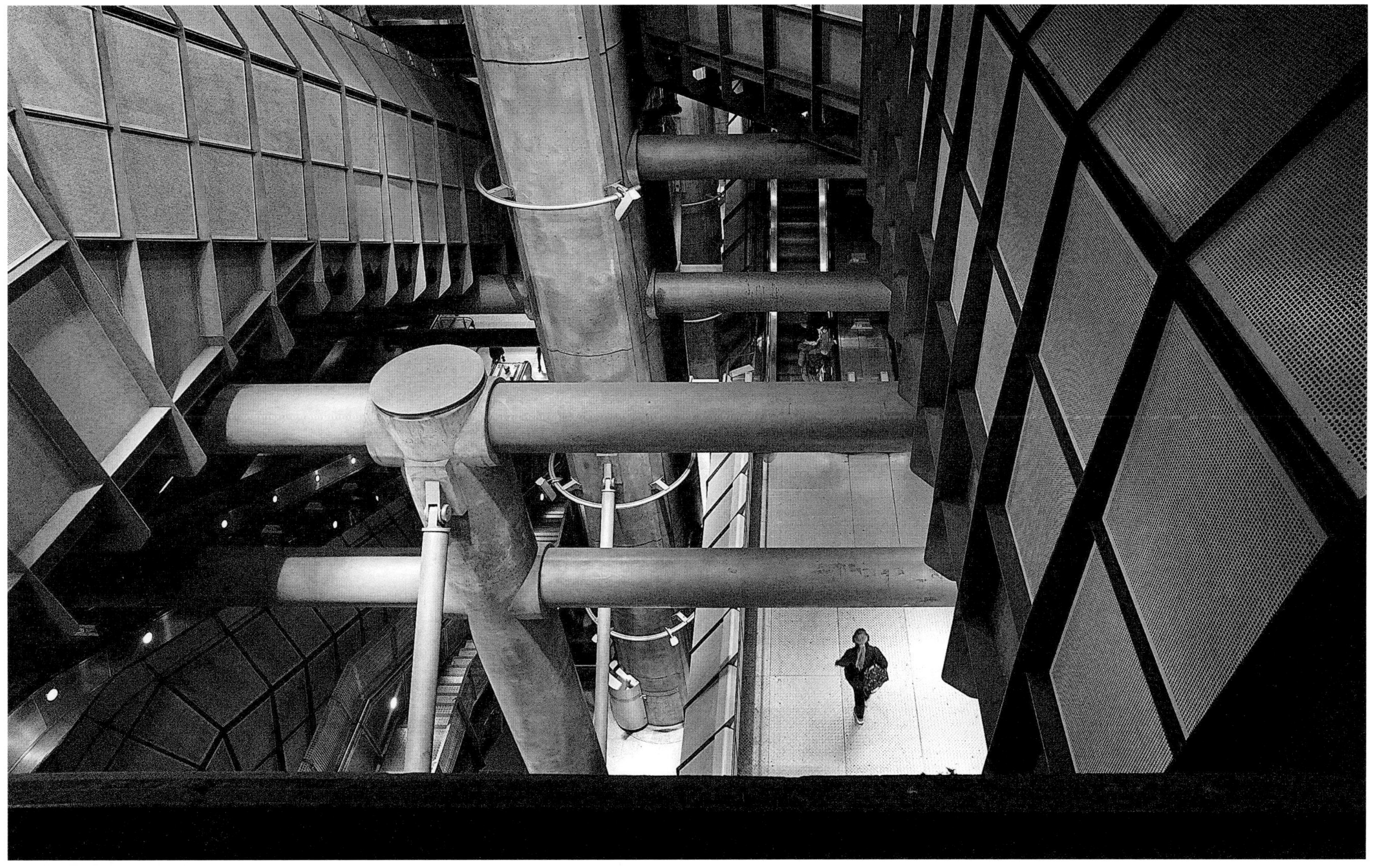

Dystopian City – Westminster

I always get *Blade Runner* vibes when I'm wandering and wondering about the striking architecture in the Jubilee line extension at Westminster, which was designed by architects Michael Hopkins and Partners, and completed in 1999.

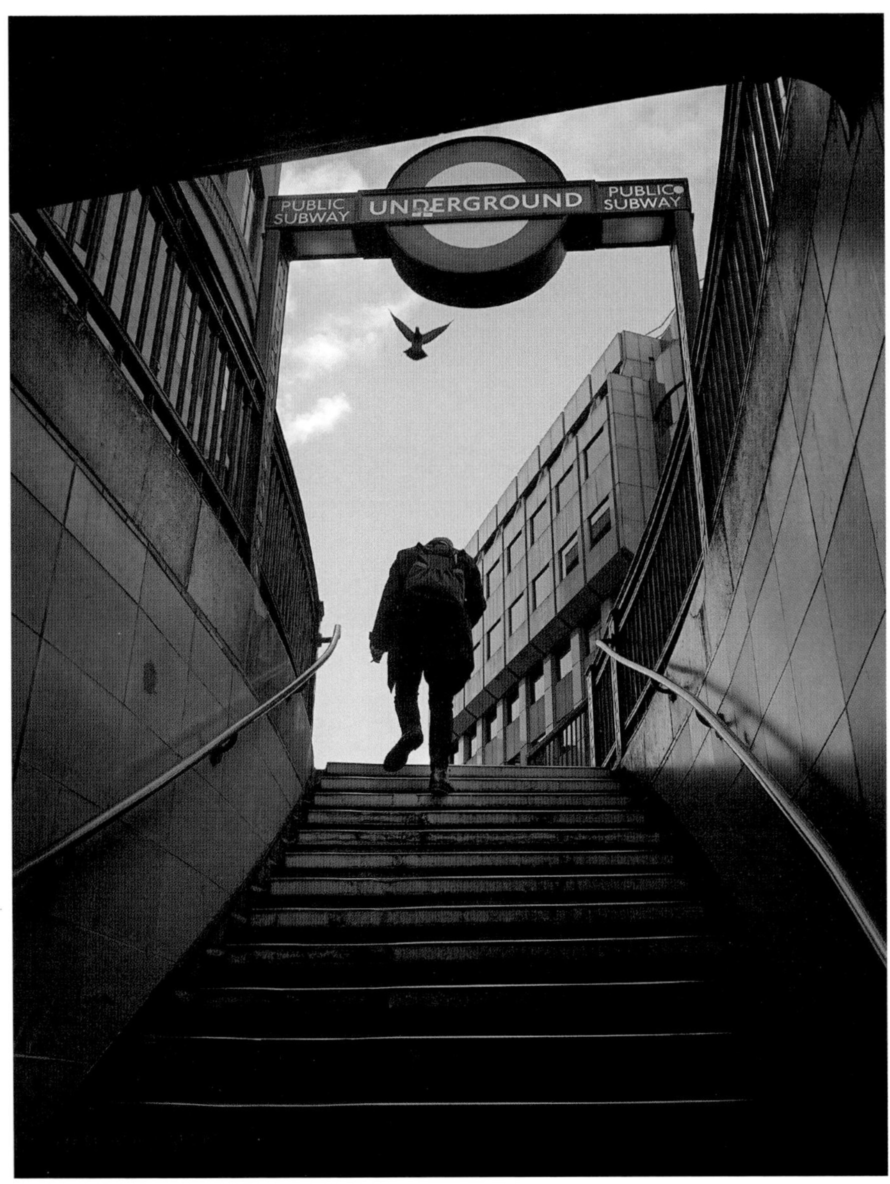

Monumental Stairs – Monument

I'm a big fan of the subway entrances to some London Underground stations. Here at Monument station, a pigeon posed for me as I captured a pedestrian climbing the stairs.

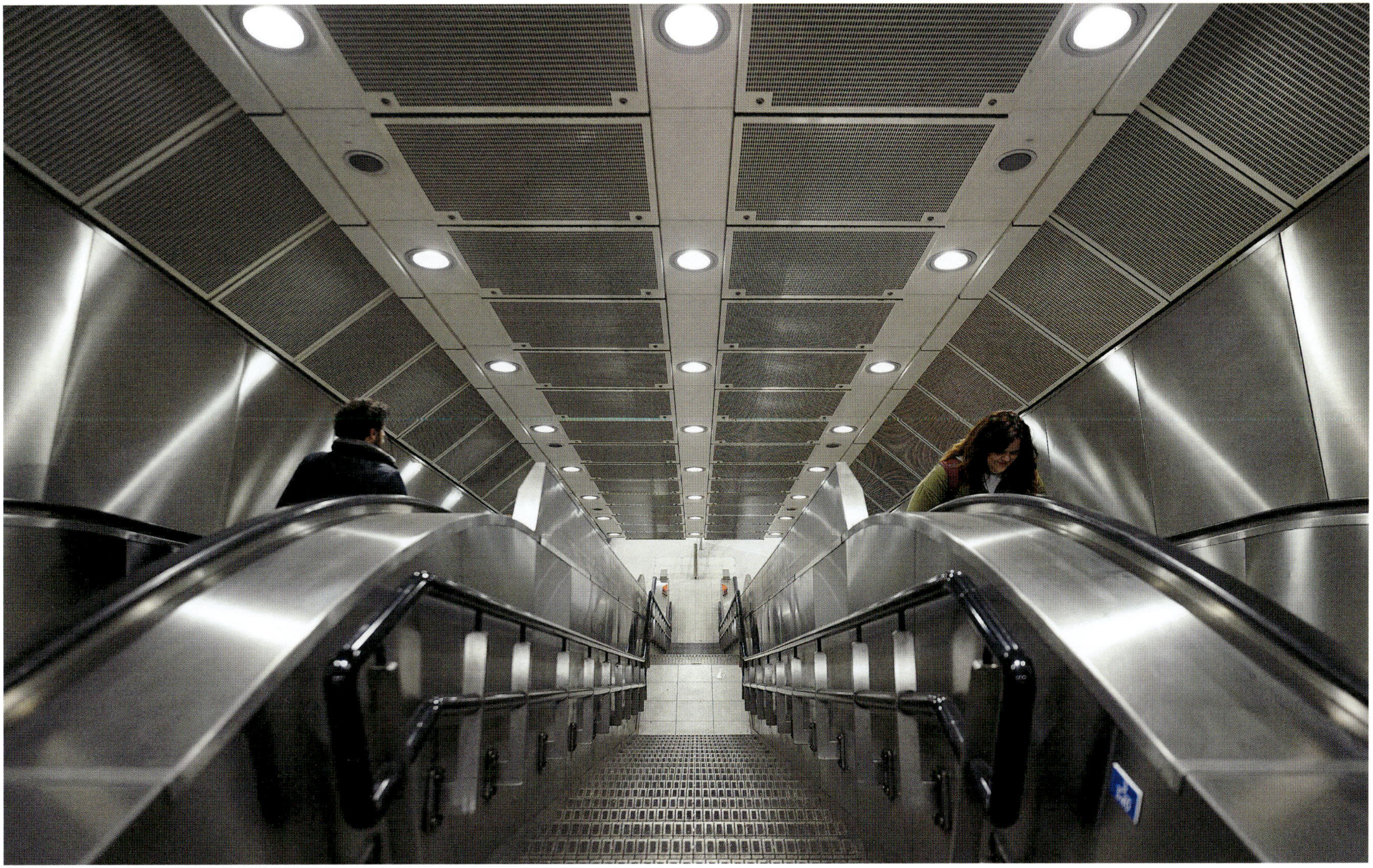

Shiny Escalators – Paddington
I've only recently discovered these shiny futuristic-looking escalators that join Paddington's Elizabeth line to the London Underground station.
They were designed by Costain Skanska Joint Venture.

Train Departs – Camden Town
If you look through the gaps in a walkway bridge above a platform at Camden Town
station, you are rewarded with an angle of this station that many have never seen.

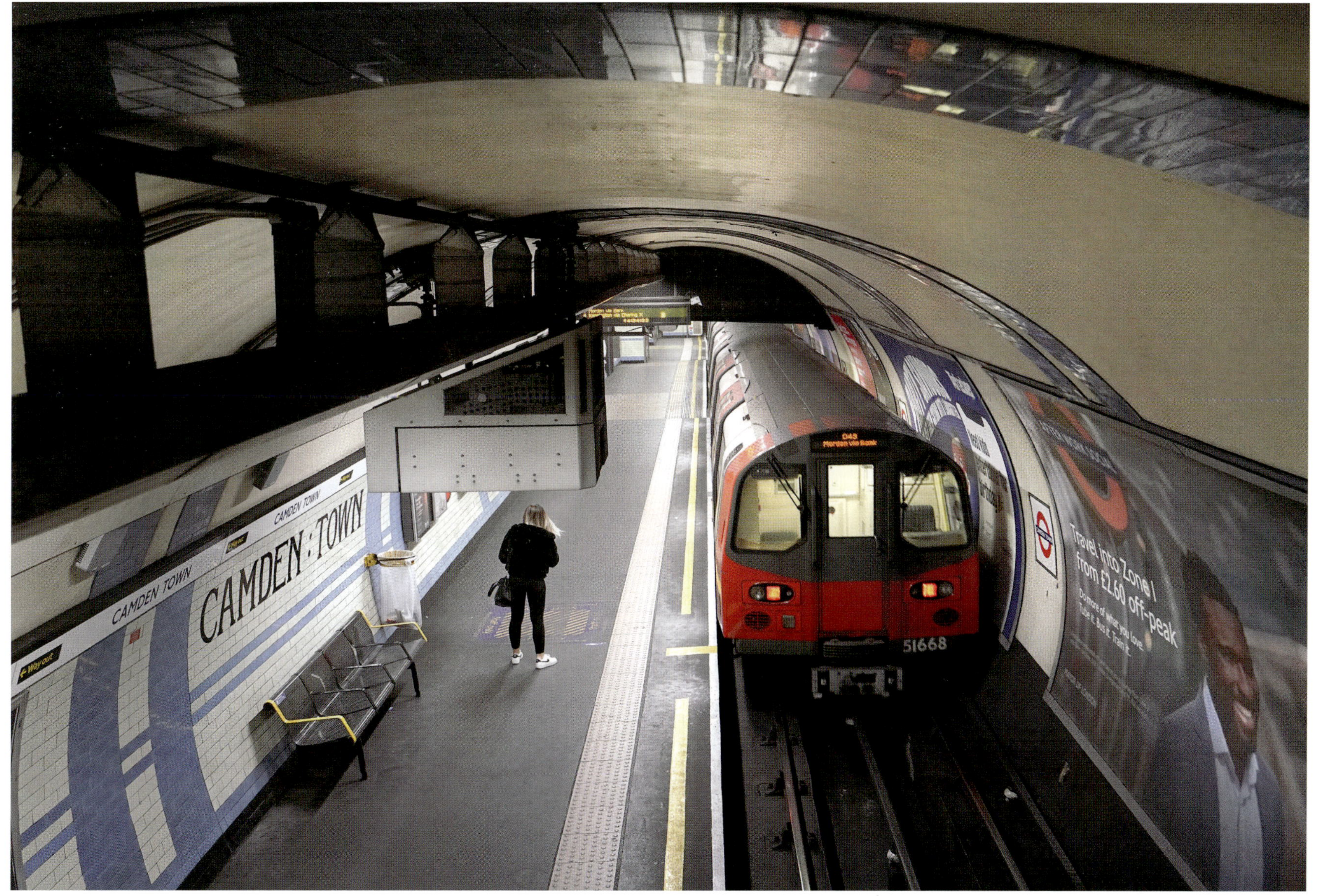
CAMDEN TOWN
CAMDEN : TOWN
CAMDEN TOWN
← Way out
Morden via Bank
Kennington via Charing X
043
Morden via Bank
51668
Travel into Zone 1
from £2.60 off-peak
Do more of what you love
Tube it. Bus it. Tram it.

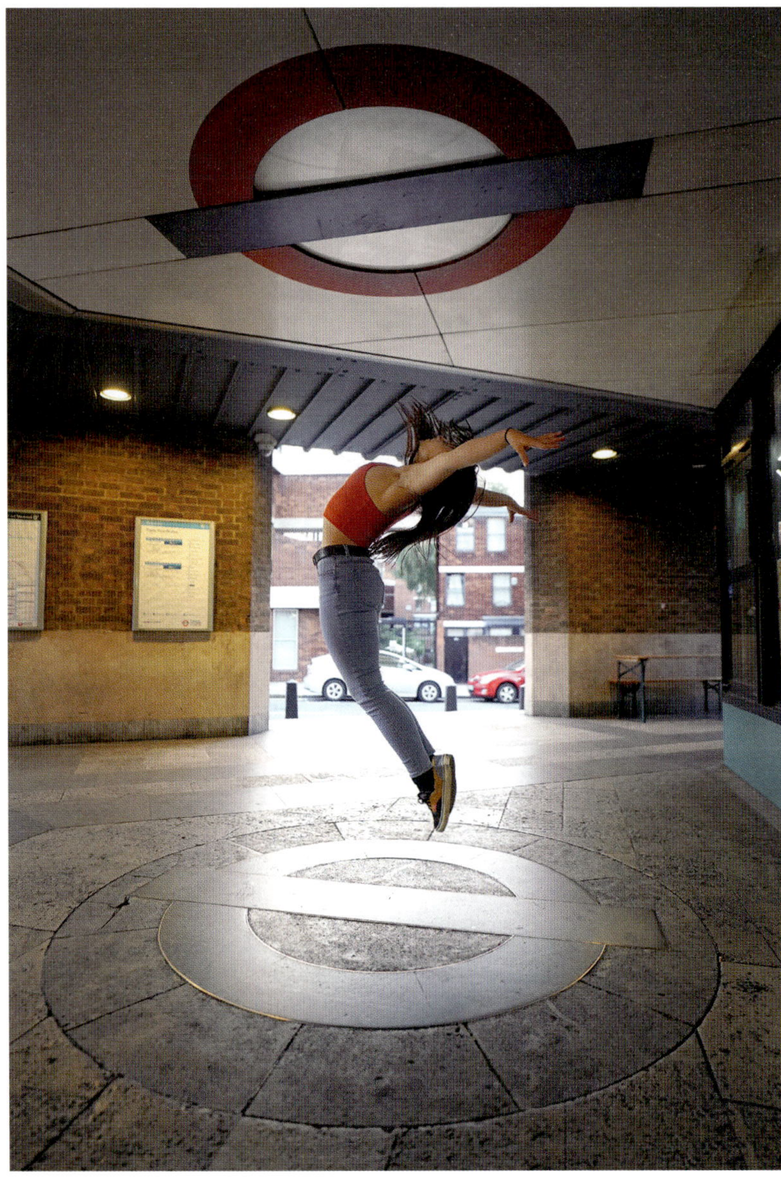

Teleportation – Pimlico
Rumour has it that if you stand on the roundel outside Pimlico station and say the name of any London Underground station three times in a row, you'll then be teleported instantly to that location.

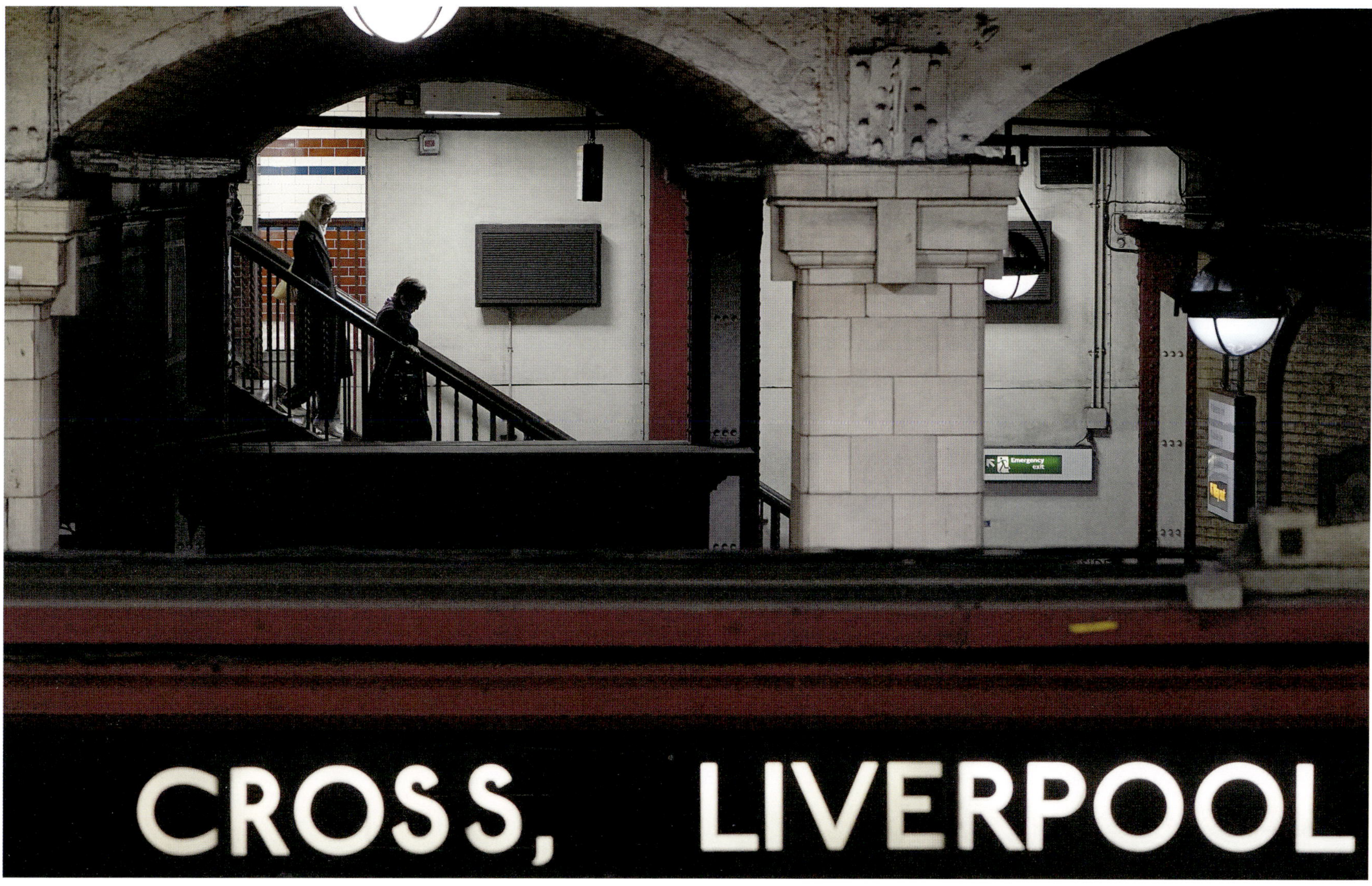

Across the Platform – Baker Street
A random, but visually pleasing, snapshot of Baker Street station, focusing on the lovely arches from across the platform.

Textures – Southwark
A close study of the reflective walls at the striking Southwark station, which was designed by
Sir Richard MacCormac and opened on 20 November 1999.

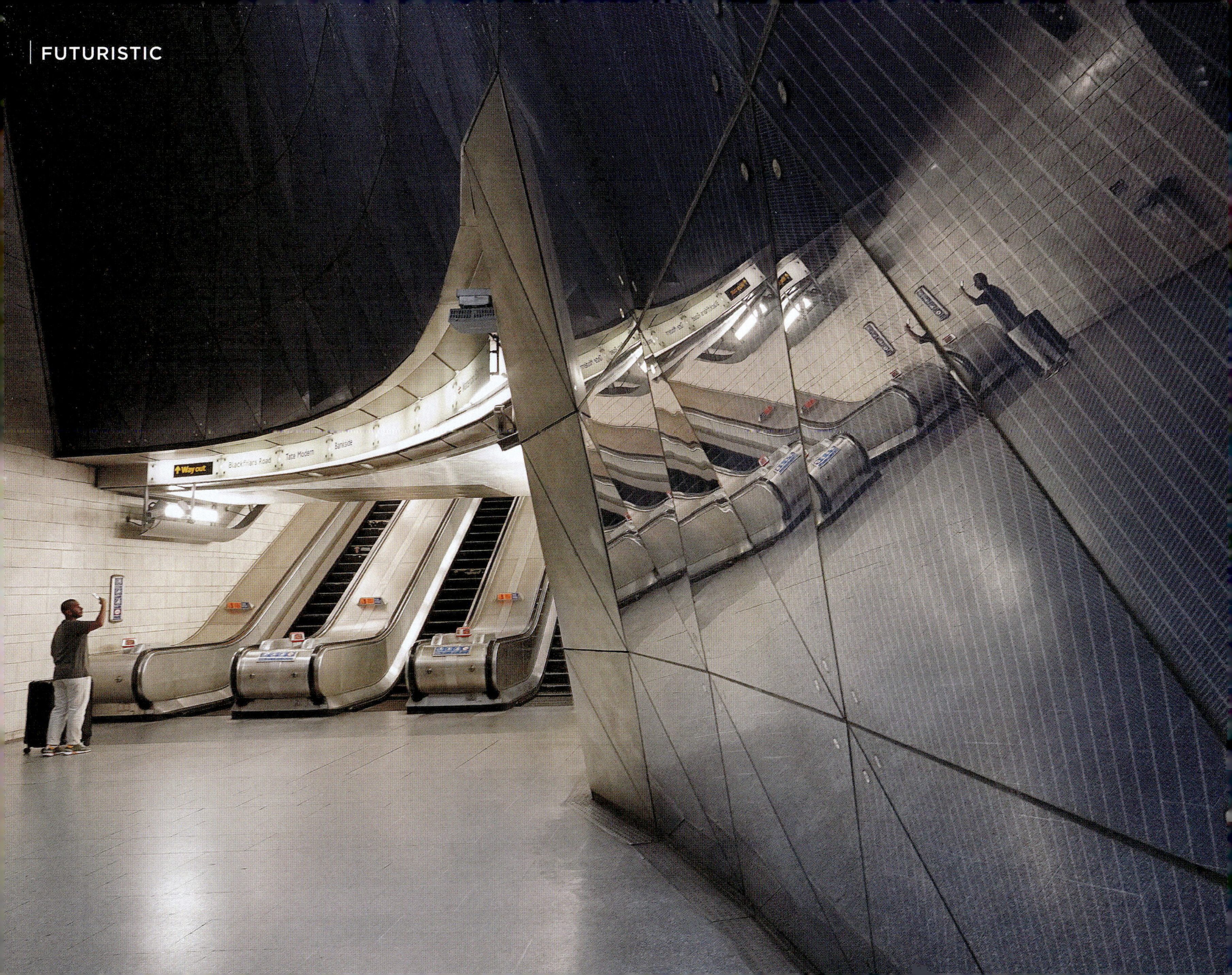
Way out
Blackfriars Road   Tate Modern   Bankside

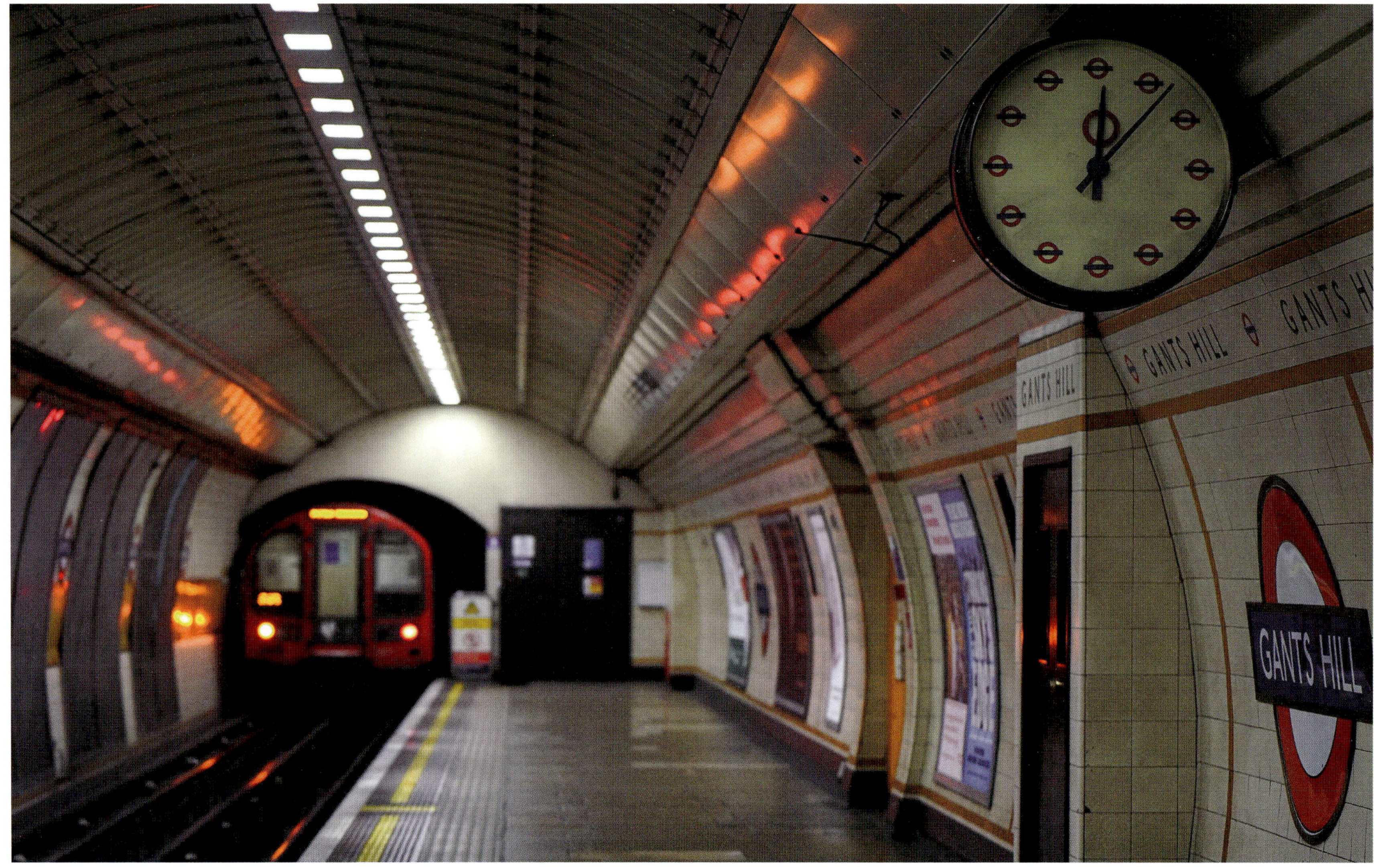

Time Travel – Gants Hill
This elegant roundel clock at Gants Hill station was designed by Magneta Time Company in the late 1940s. I love the reflections in the background created by the departing train.

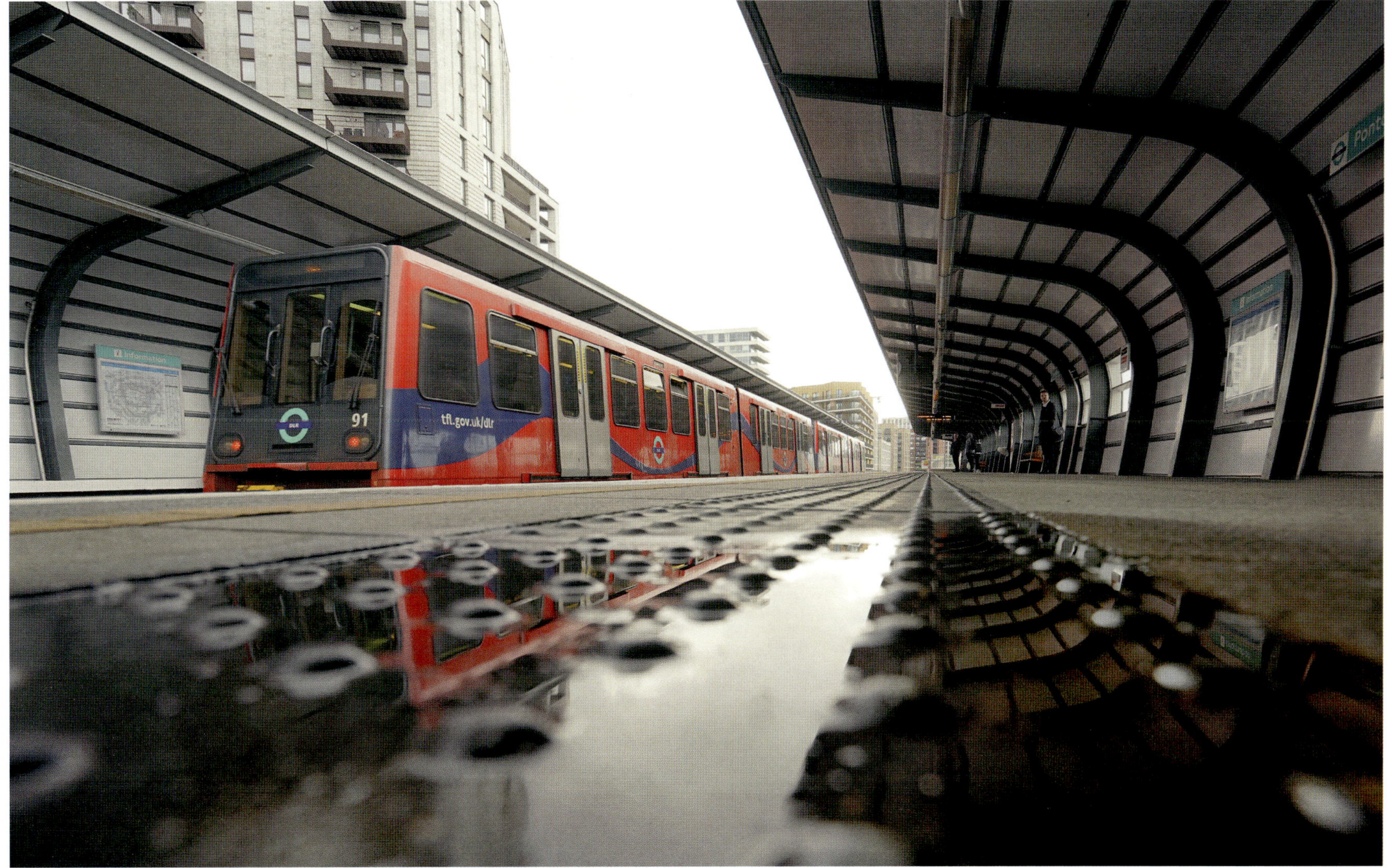

Puddle Reflections – Pontoon Dock
A train departs the sleek and elegant DLR platform. I'm very fond of the curved canopy cover design by architects
Weston Williamson + Partners.

Stairs – Covent Garden
A woman in a stylish coat goes deeper underground at Covent Garden station.
This station has lovely tiling and a fifteen-storey spiral staircase.

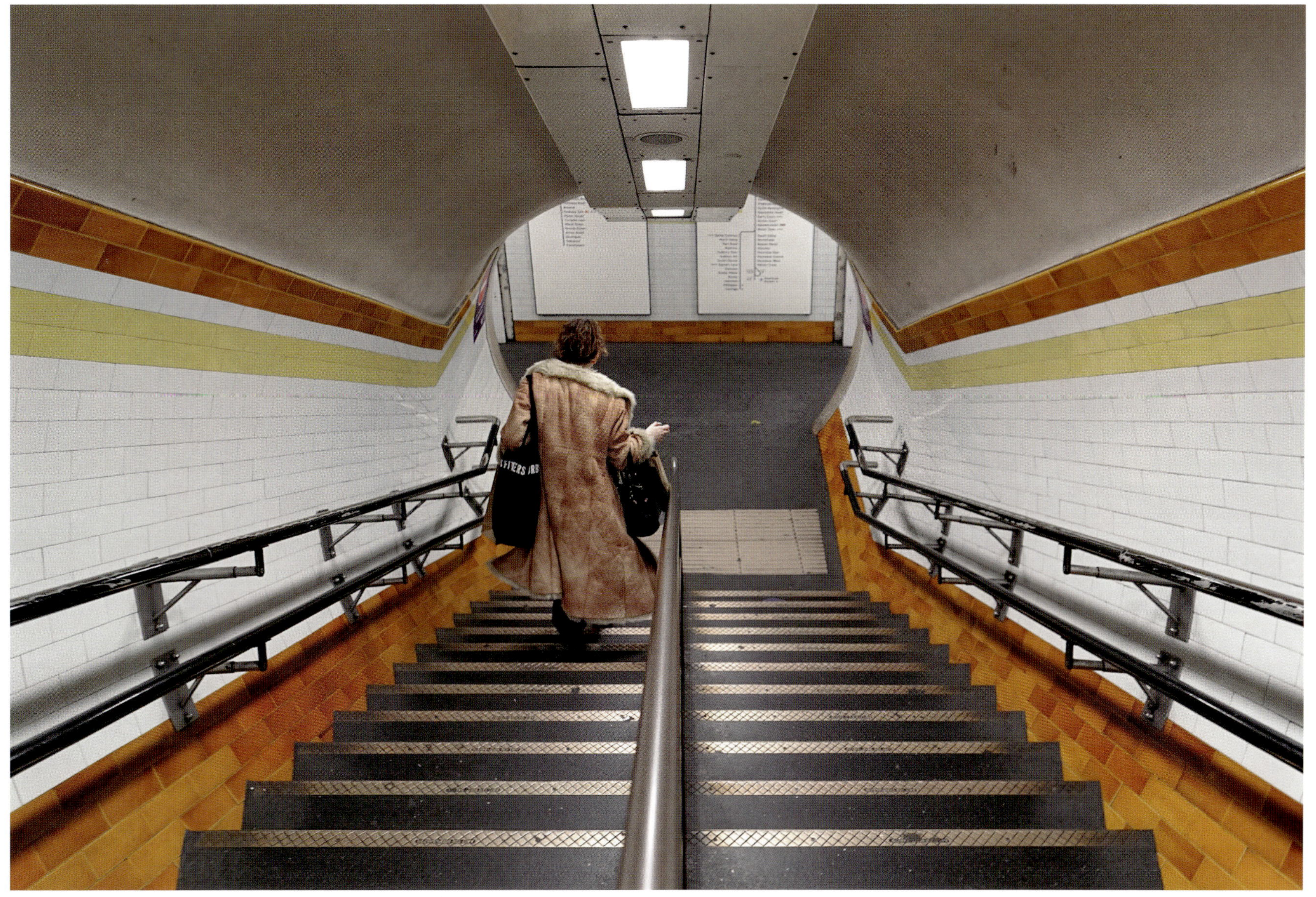

Centre Stage – London Bridge
Stepping into the Jubilee line extension at London Bridge station feels like you've immediately arrived in the future, as the hallways and tunnels resemble the interior of a spaceship. I got lucky here, as the pedestrian's clothes matched the surrounding colours.

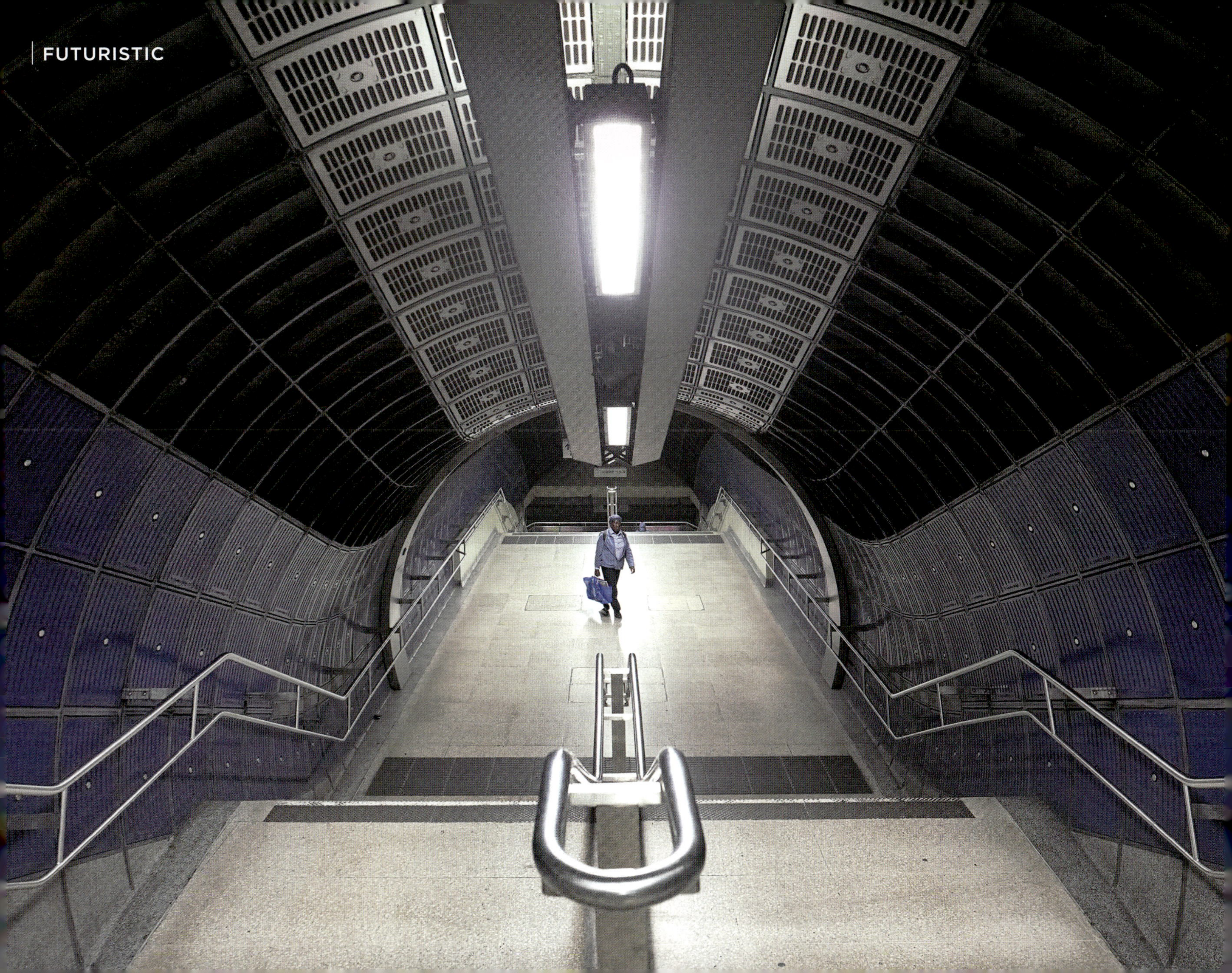

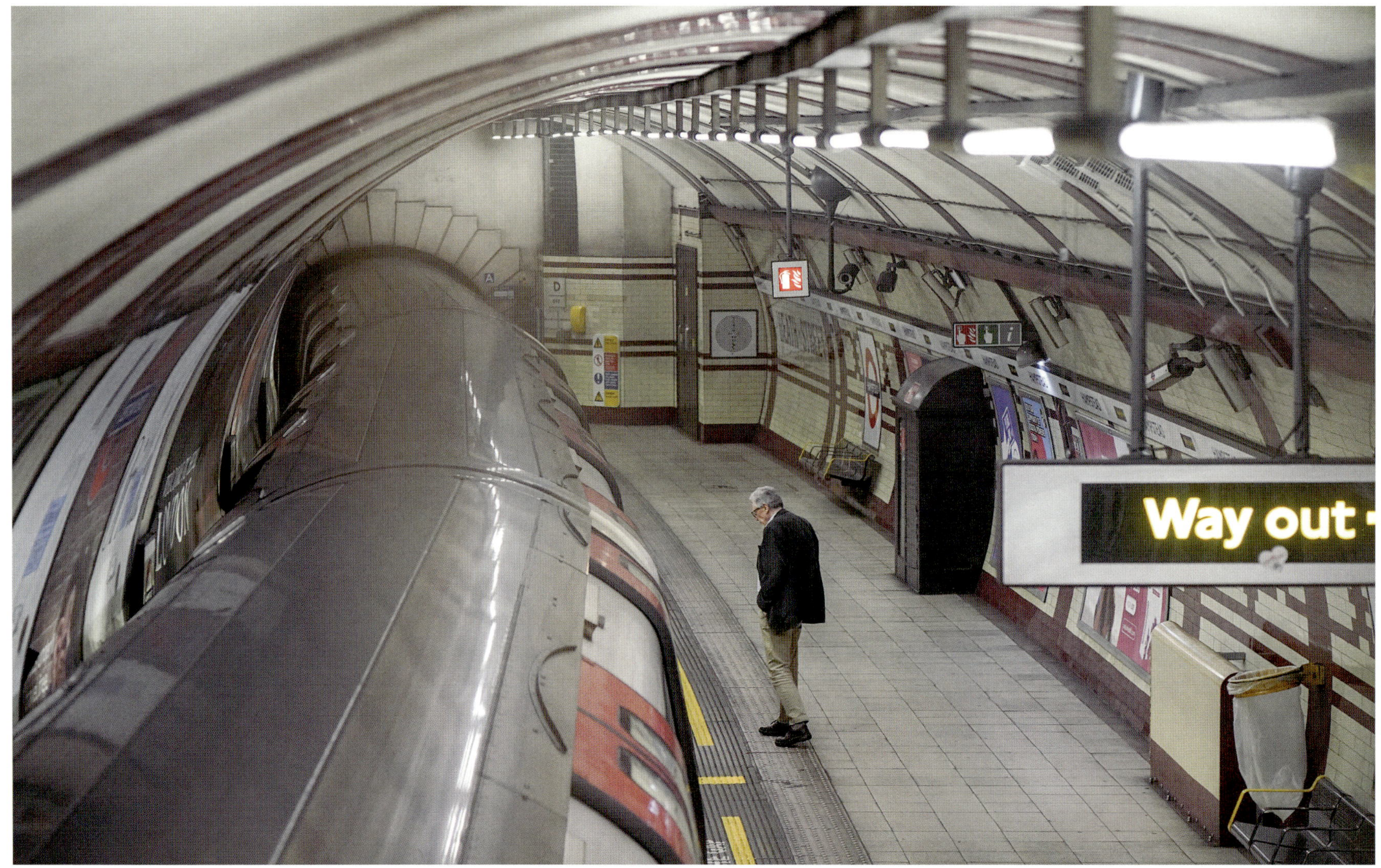

Waiting – Hampstead
A unique point of view, looking down on a traveller waiting for a train arriving at Hampstead station, surrounded by beautiful tiles and architecture.

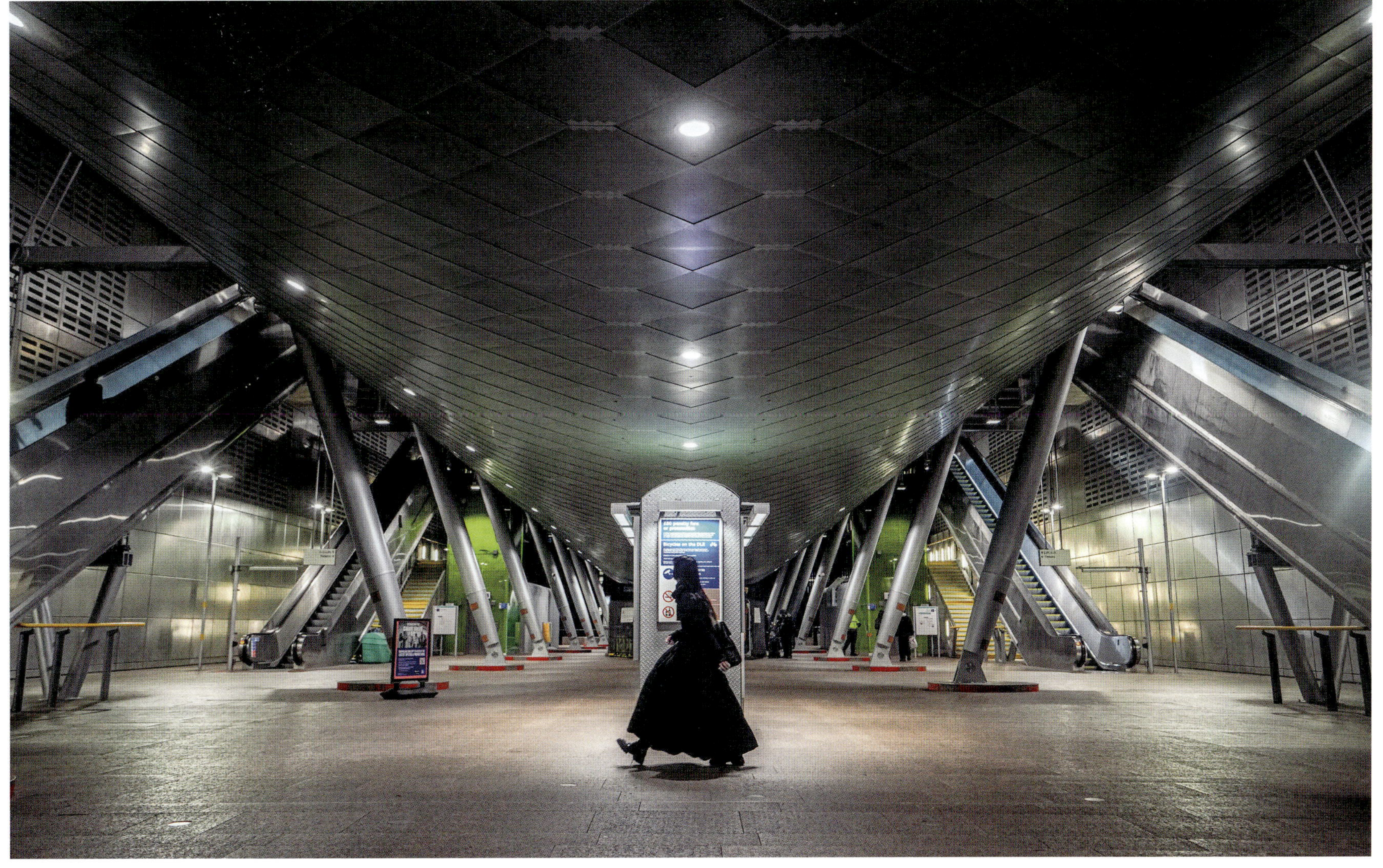

Under the Spaceship – Heron Quays

A lady walks beneath what appears to be the underside of a spaceship. A few people have said the lady's long garments made them think of a Jedi knight or Sith lord from *Star Wars*.

To Finsbury Park – Hyde Park Corner
Beautiful old signage and tiles are on the platform at Hyde Park Corner station, which opened on
5 December 1906. I love the little details you can discover when exploring stations.

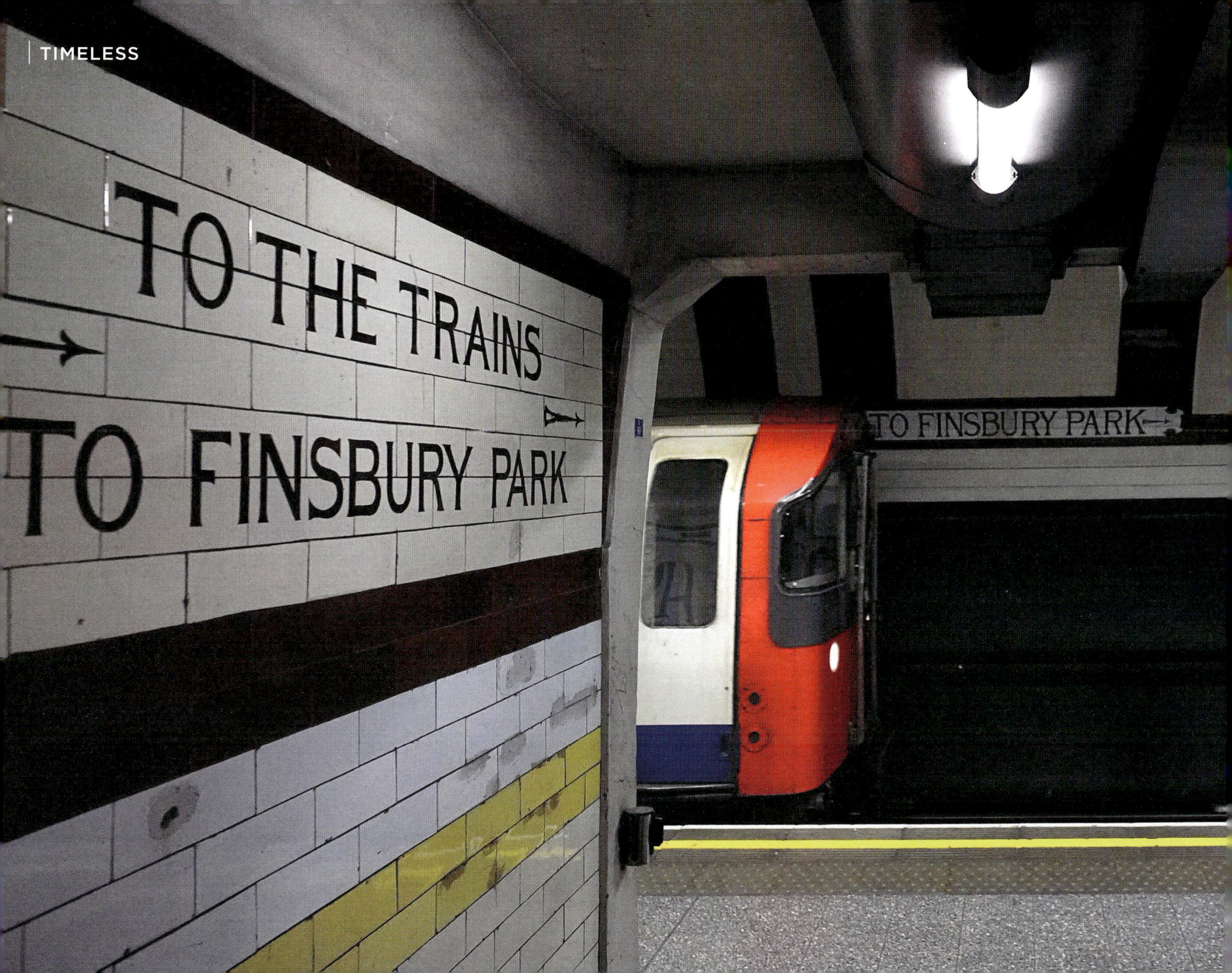
TO THE TRAINS
TO FINSBURY PARK
TO FINSBURY PARK

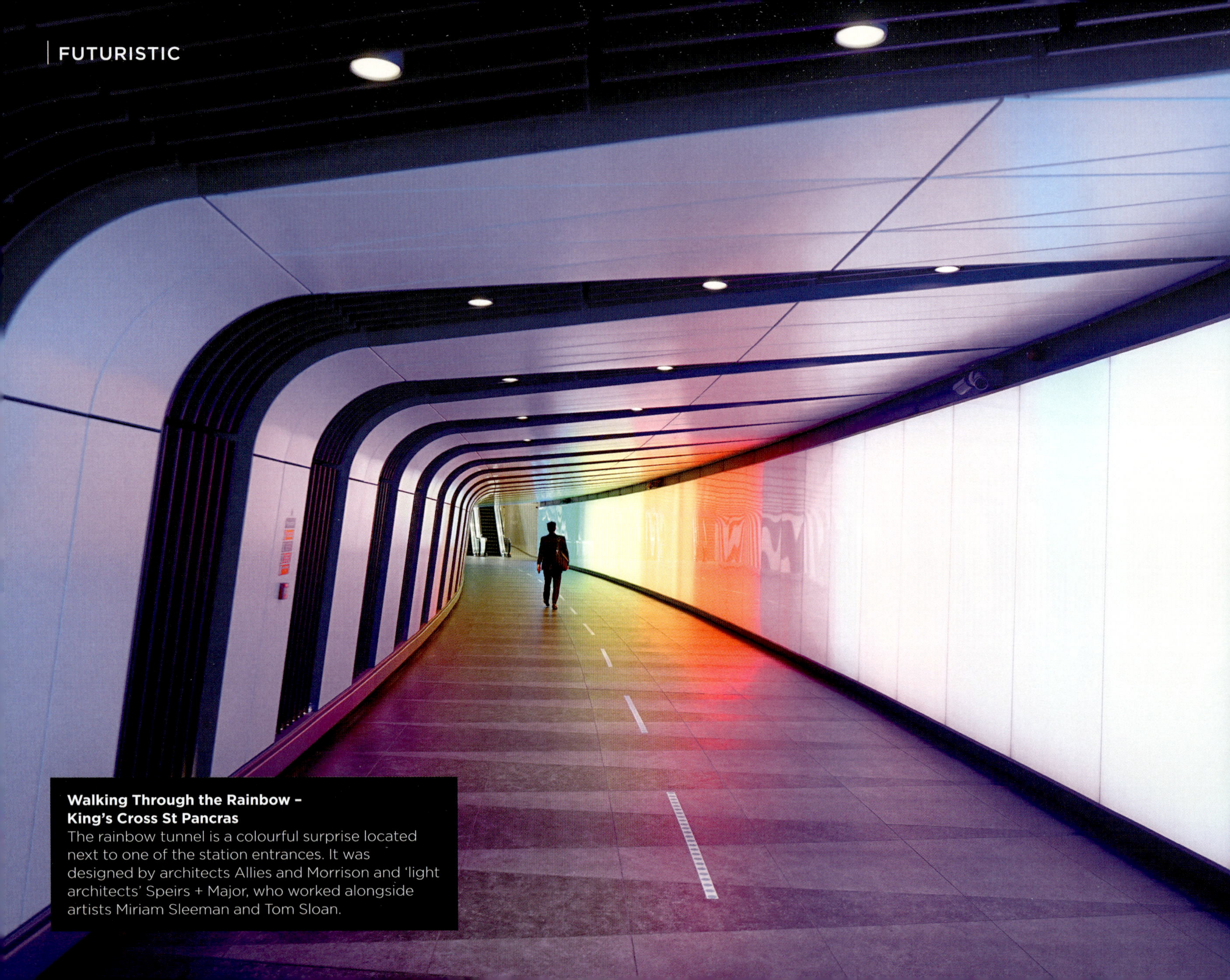

**Walking Through the Rainbow –
King's Cross St Pancras**
The rainbow tunnel is a colourful surprise located next to one of the station entrances. It was designed by architects Allies and Morrison and 'light architects' Speirs + Major, who worked alongside artists Miriam Sleeman and Tom Sloan.

King William Street's Lament

As London rose up from the past,
All steel-clad bones and shining glass,
'Neath sullen streets an old way dreamed,
Its tunnels sealed and unredeemed.
'Til jealous progress burrowed through,
Devouring old to feed the new.

The station screamed, bereft of heart.
Its aged fabric torn apart.
By drills that cared not what they ground,
So long as profit could be found.
A relic lost, its fate foregone
And progress marches blindly on.

Matthew Ward

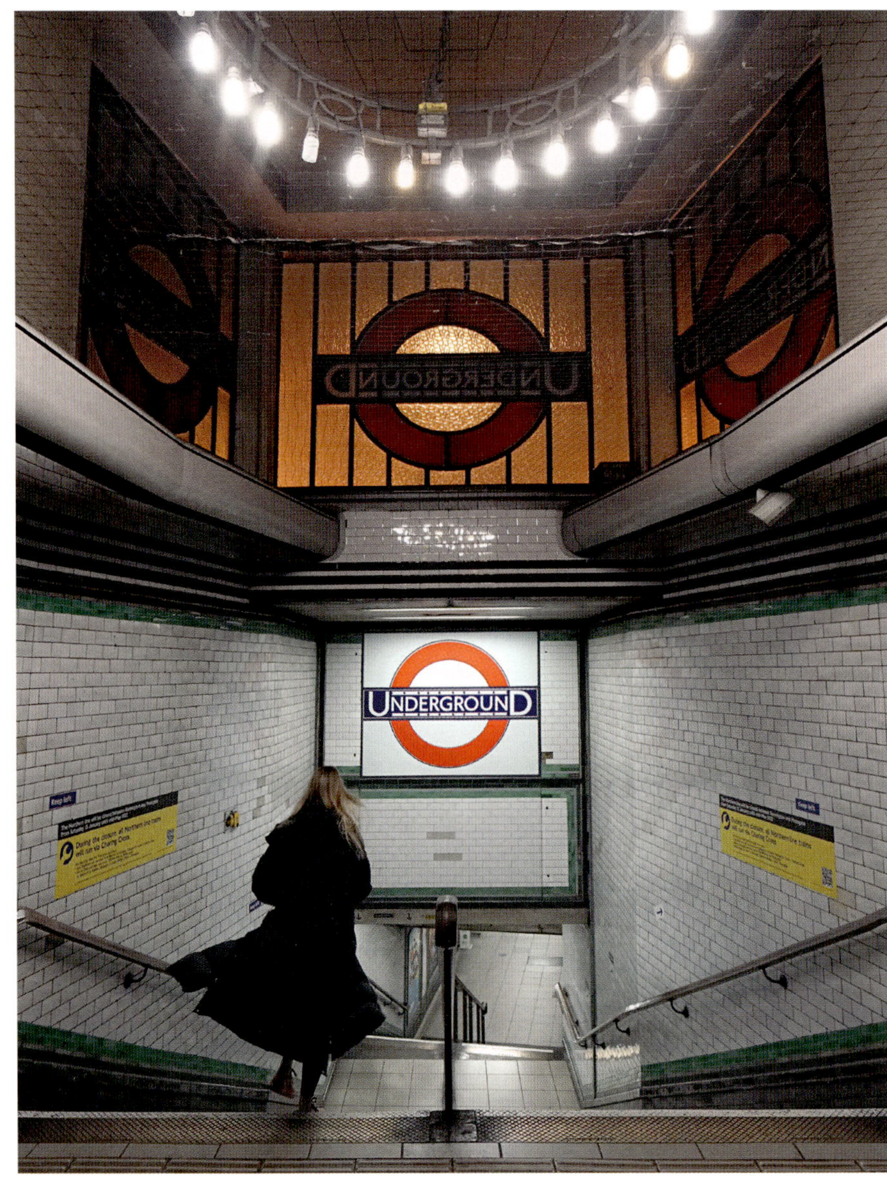

Coats, Stairs and Chandeliers – Tooting Bec
A lady rushes to catch her train, surrounded by the stunning architecture
of Charles Holden. Tooting Bec station opened on 13 September 1926.

3

NIGHT LIGHT/DAY LIGHT

One of the aspects that I most enjoy about living in London is that there is always something happening somewhere. This notion is also true about the Underground. The service isn't 24/7 like the New York Subway (although, on 19 August 2016, TfL introduced a Night Tube service, which means there are trains running twenty-four hours on Fridays and Saturdays), but there is constant work going on in the control room, maintaining, cleaning and upgrading the service. Also, the city has people working at all hours, meaning there is generally activity somewhere on the network.

The London Underground is a visual spectacle no matter what time you travel. During the day, I like observing how the shadows and sunlight interact with the architecture, introducing light and highlighting features that are easy to disregard. Meanwhile, the city takes on a new life at night, like it's wearing its 'going out' clothing. The artificial lights welcome new colours to architecture, giving structures new meaning in their transformed appearance. Heavy rainfall further enhances this nocturnal transformation.

Photo Hints

Night Light and *Day Light*: Spend time at locations and watch how the light interacts with the architectural elements of the stations. Observe how this changes as the sun moves throughout the day or when artificial light is added during the evening. Timing really affects composition here.

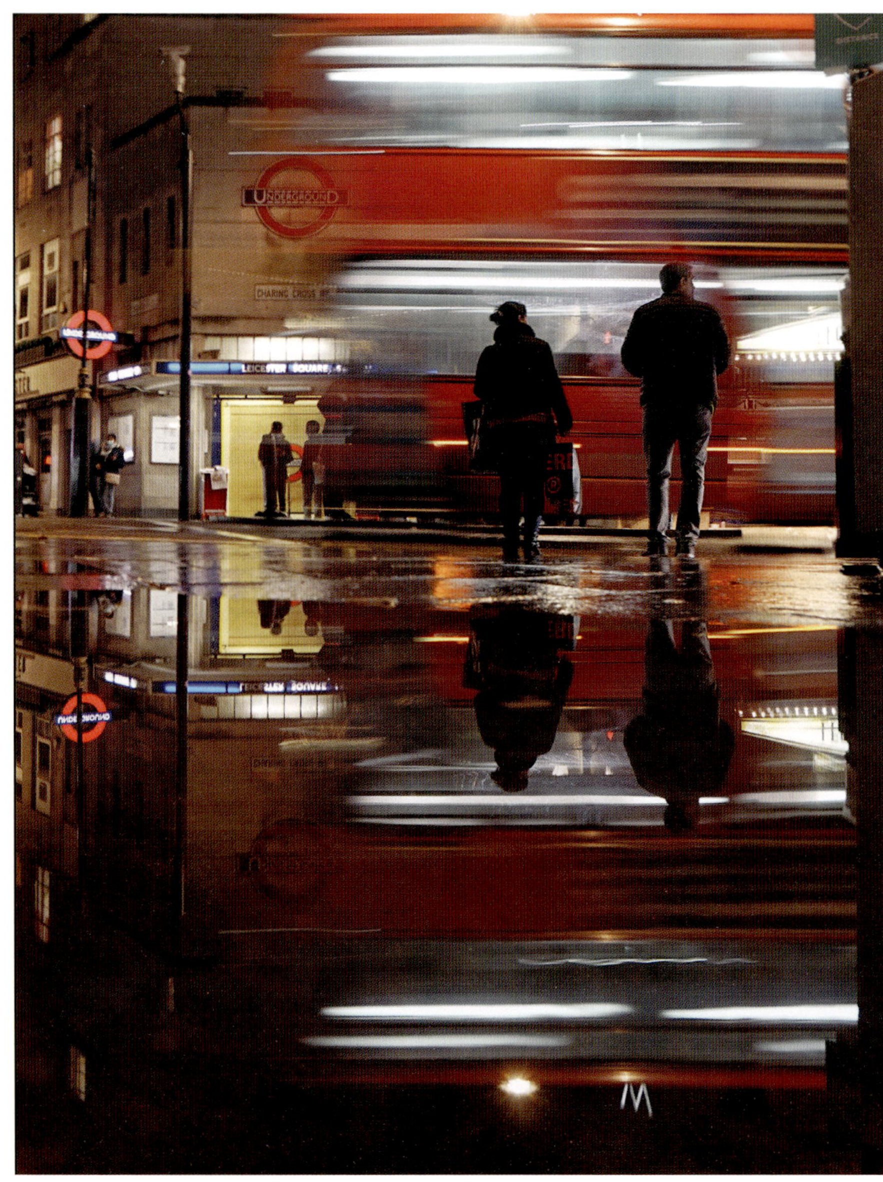

Waiting – Leicester Square
I love the variety of colours on show as a bus zooms past Leicester Square station on a rainy night. My wife thinks this image looks like a painting!

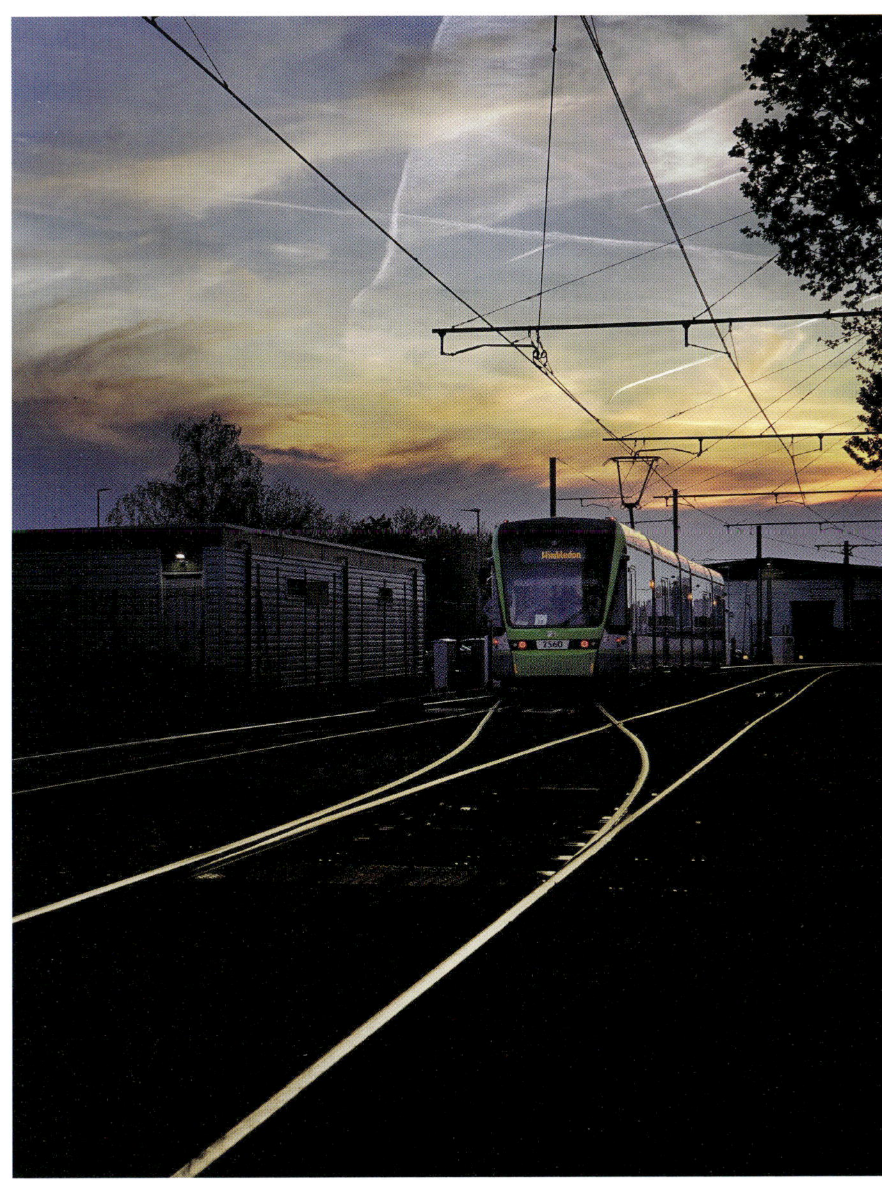

Sunset – Therapia Lane
I love how photogenic the Croydon tram network is. Whenever I capture photography of the route, I feel like I've discovered a new secret London Underground line. Here's a shot of a tram arriving at Therapia Lane stop as the sun sets.

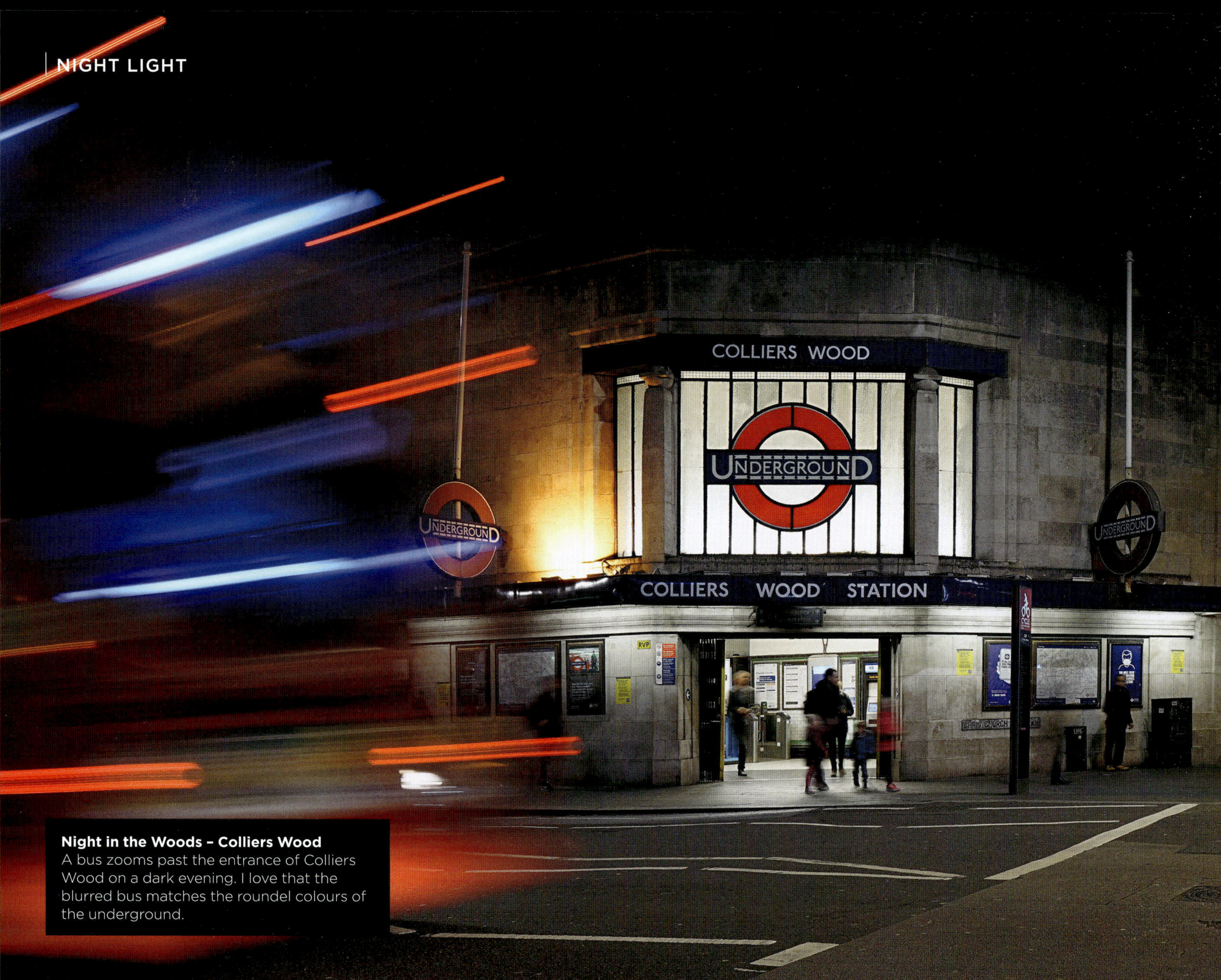

Night in the Woods – Colliers Wood
A bus zooms past the entrance of Colliers Wood on a dark evening. I love that the blurred bus matches the roundel colours of the underground.

City at Night

Can't see the stars down there,
too much light to see their light,
piercing the darkness of midnight.
Yet, you know that they are there,
like the intent of city schemers,
hidden among late night revellers.
Light hides the darkness of light,
while the economic machine dazzles
hiding blankets bright and frazzled.
Can't see the stars down there.

Danny Herbert

Sun Angel – Prince Regent

I always enjoy trying to view objects and structures differently. For example, I've noticed that the canopy covers on the platforms of some DLR stations resemble angel wings. Here's a fun photograph of dancer Andreya posing at Prince Regent DLR station.

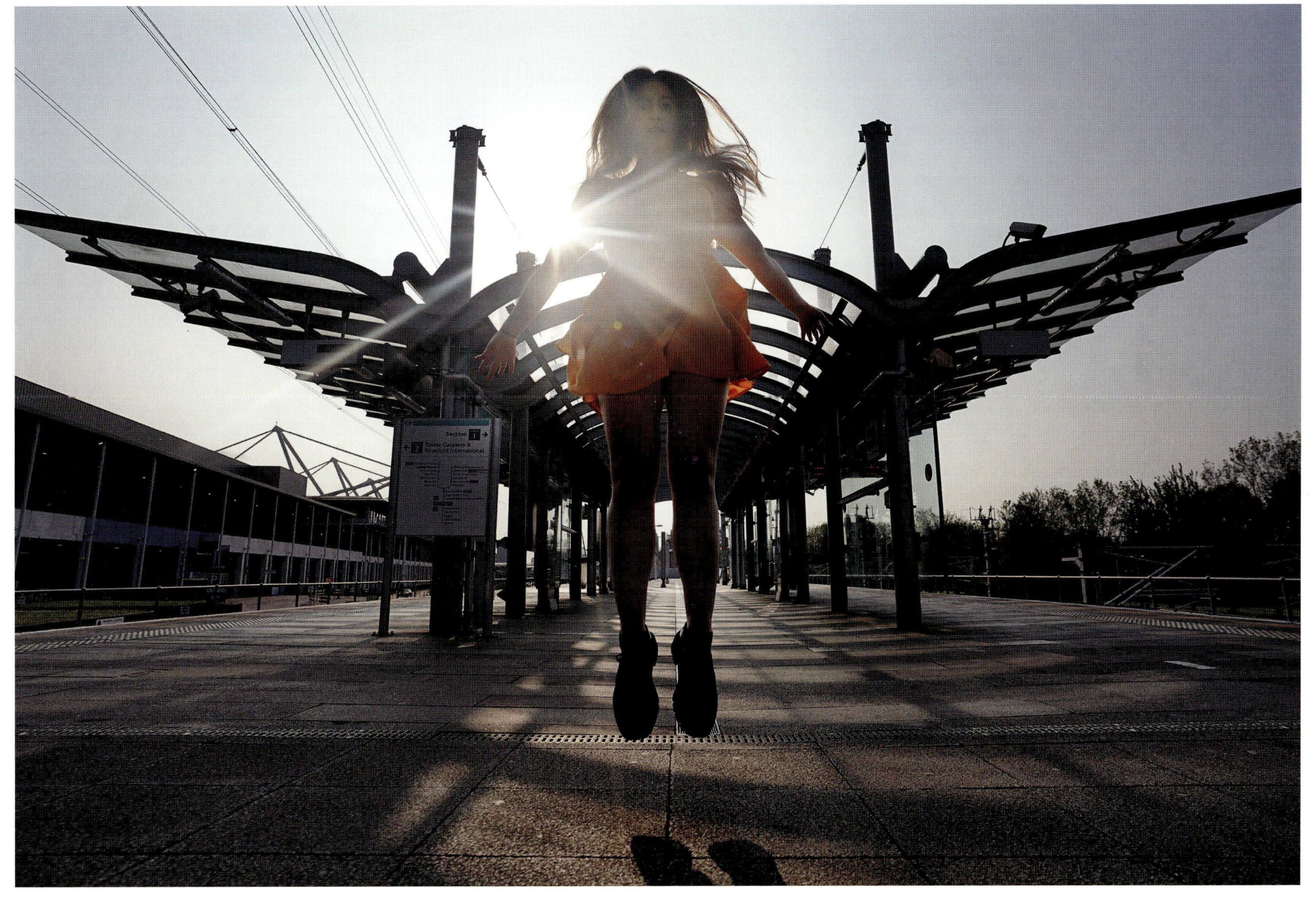

Night Reflections – Euston
I couldn't resist some puddle photography outside Euston Underground station;
I love the colourful reflections of the building lights contrasted with the deep black night sky.

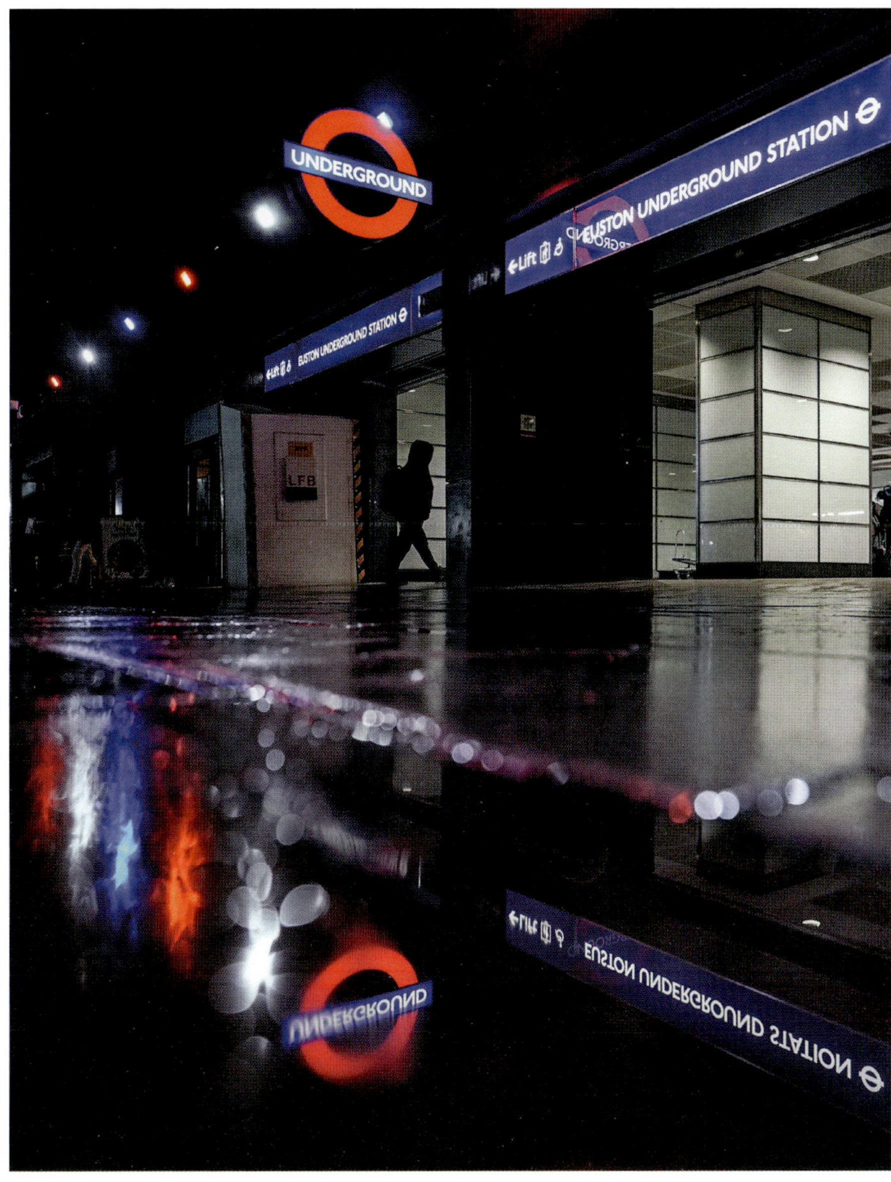

UNDERGROUND
EUSTON UNDERGROUND STATION
EUSTON UNDERGROUND STATION
← Lift
LFB

Roundel Projection – Nine Elms

The beautiful roundel projection from the window at Nine Elms station. Whenever I travel here, I always check if the sun is shining so that there's a possibility of capturing this.

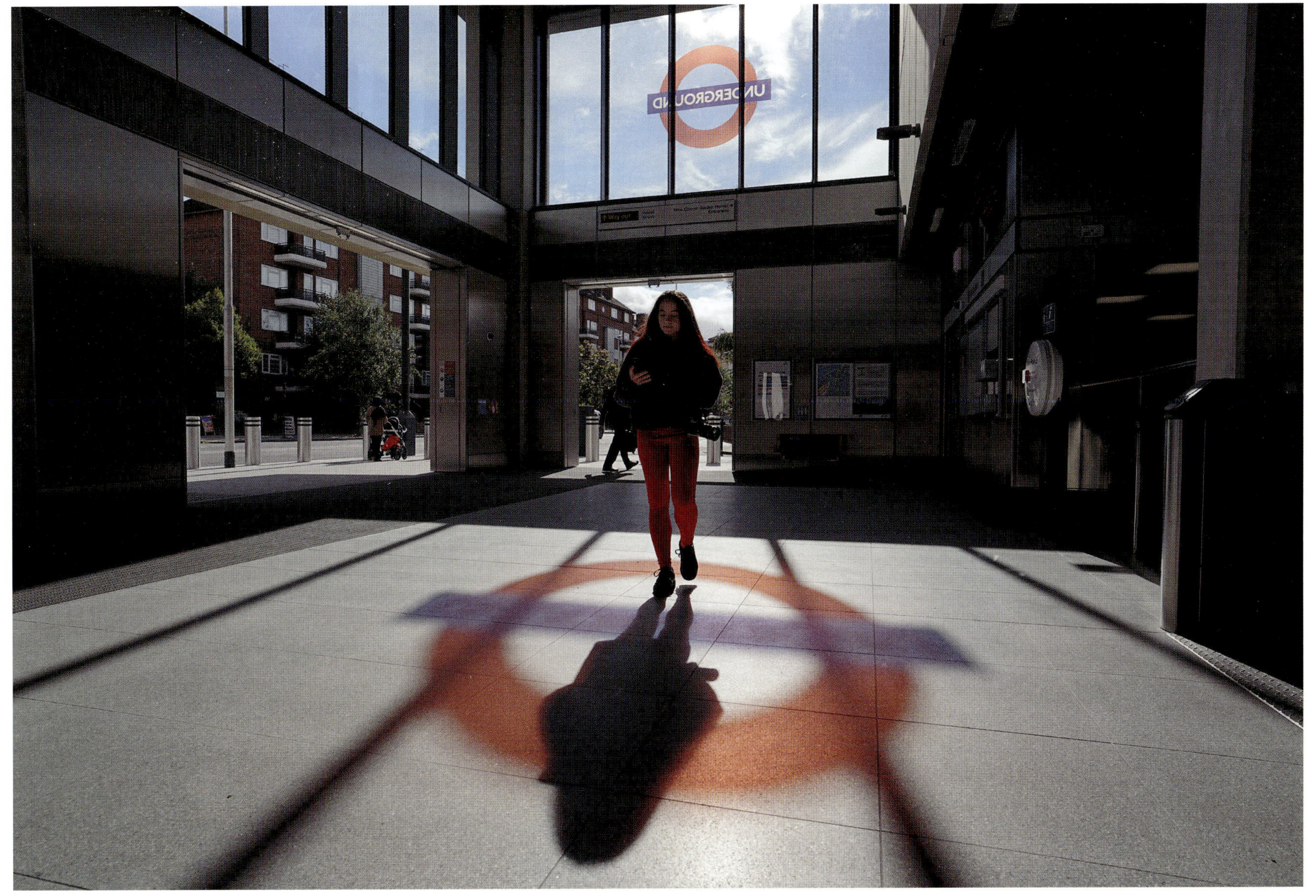

UNDERGROUND

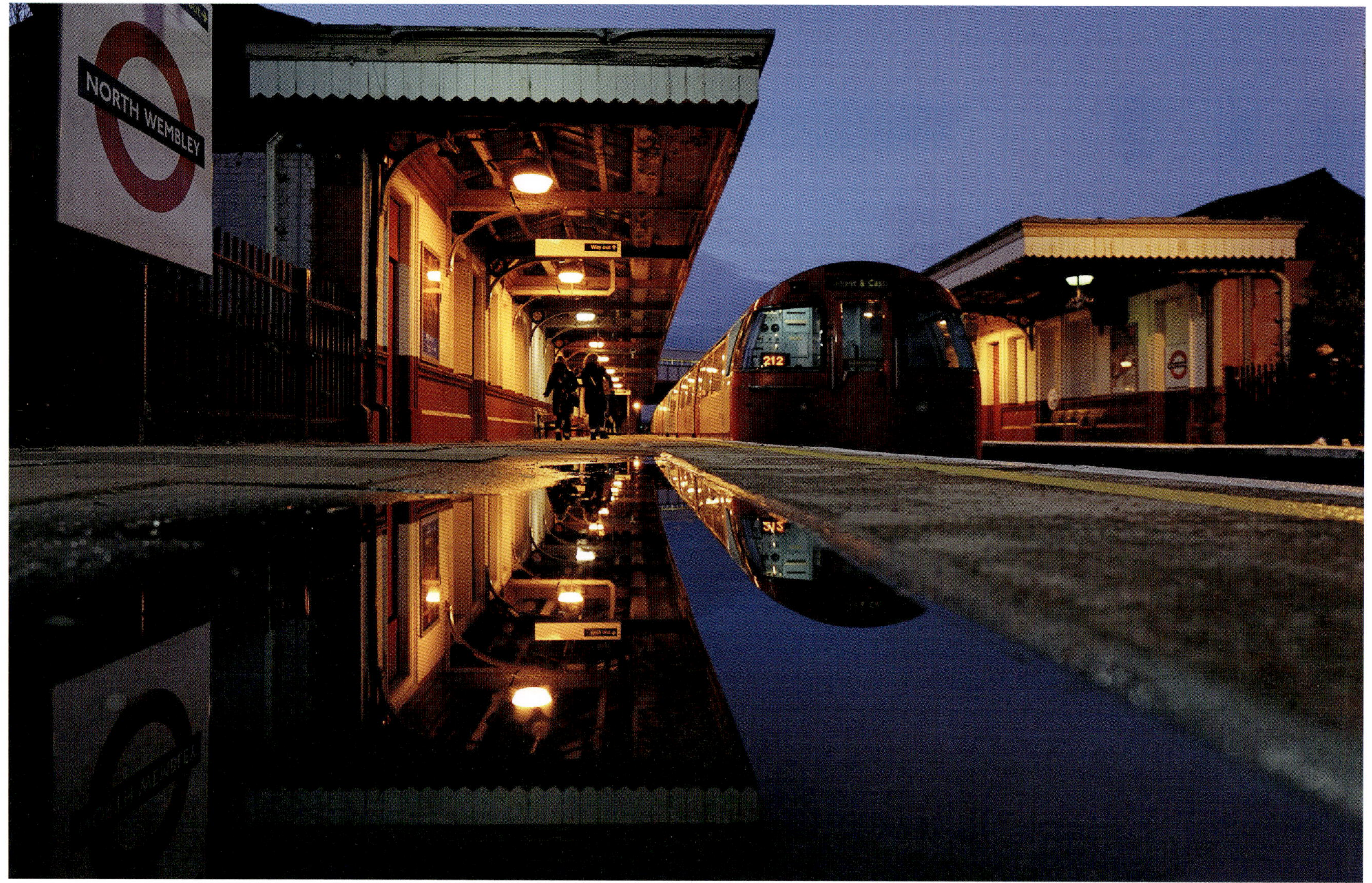

Train Departs – North Wembley
A nearby puddle reflects a train leaving North Wembley station at blue hour on a rainy evening.

winter rain
passengers carry home
droplets of night

John Hawkhead

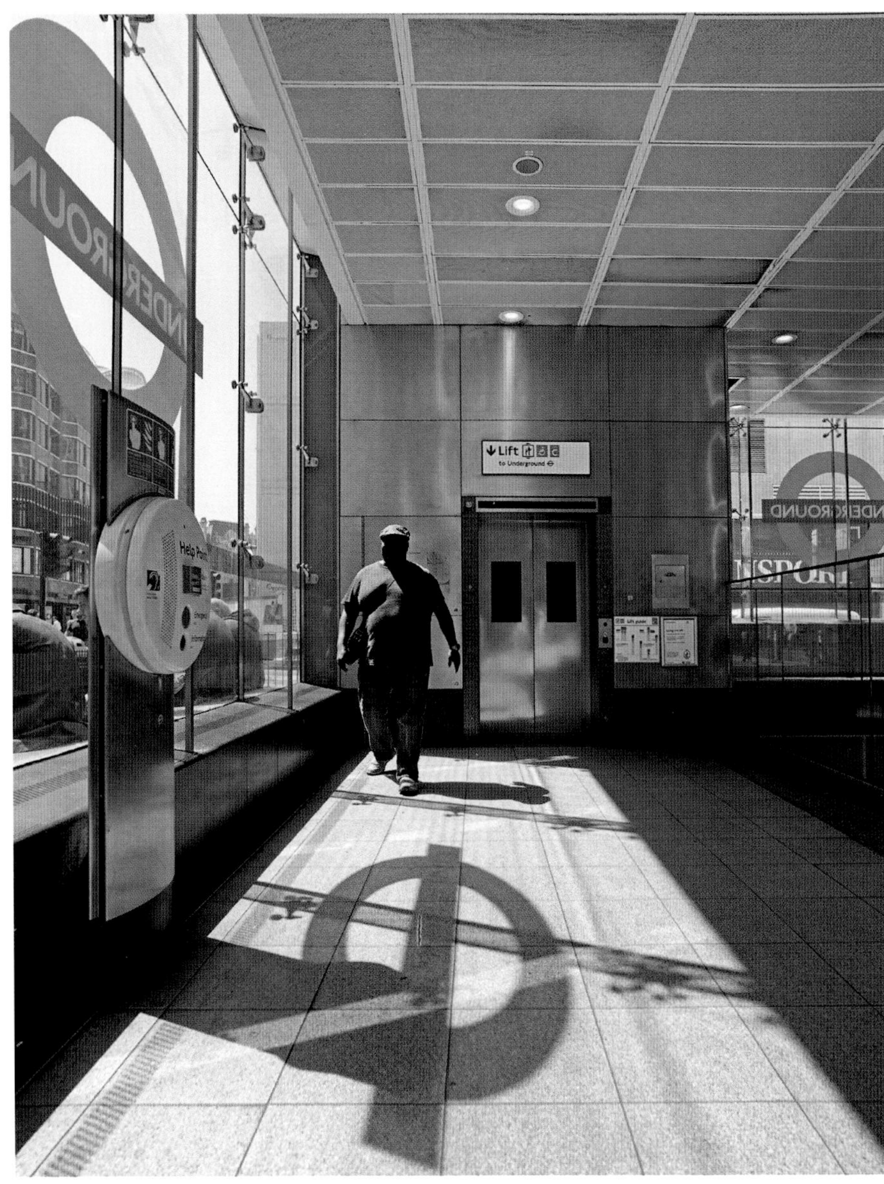

Shadows, Reflections and Projections – Victoria
The next time you pass through Victoria Underground station on a sunny day, head towards the entrance at Cardinal Place. If you're lucky, you'll be presented with the delightful sight of a roundel projected onto the floor.

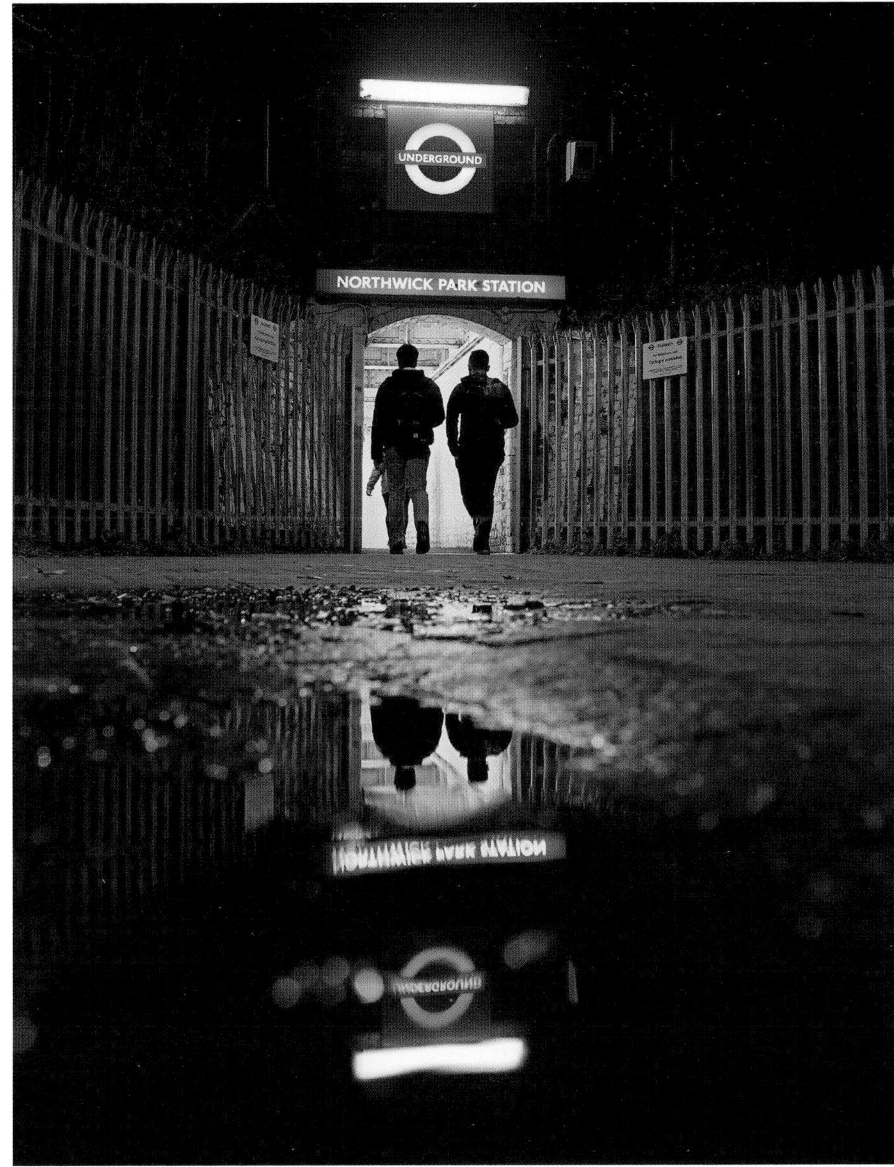

Entrance – Northwick Park
A couple of pedestrians walk towards the entrance to Northwick Park station on a rainy evening. I couldn't ignore the well-placed puddle in front of the opening; it was well worth getting wet knees for.

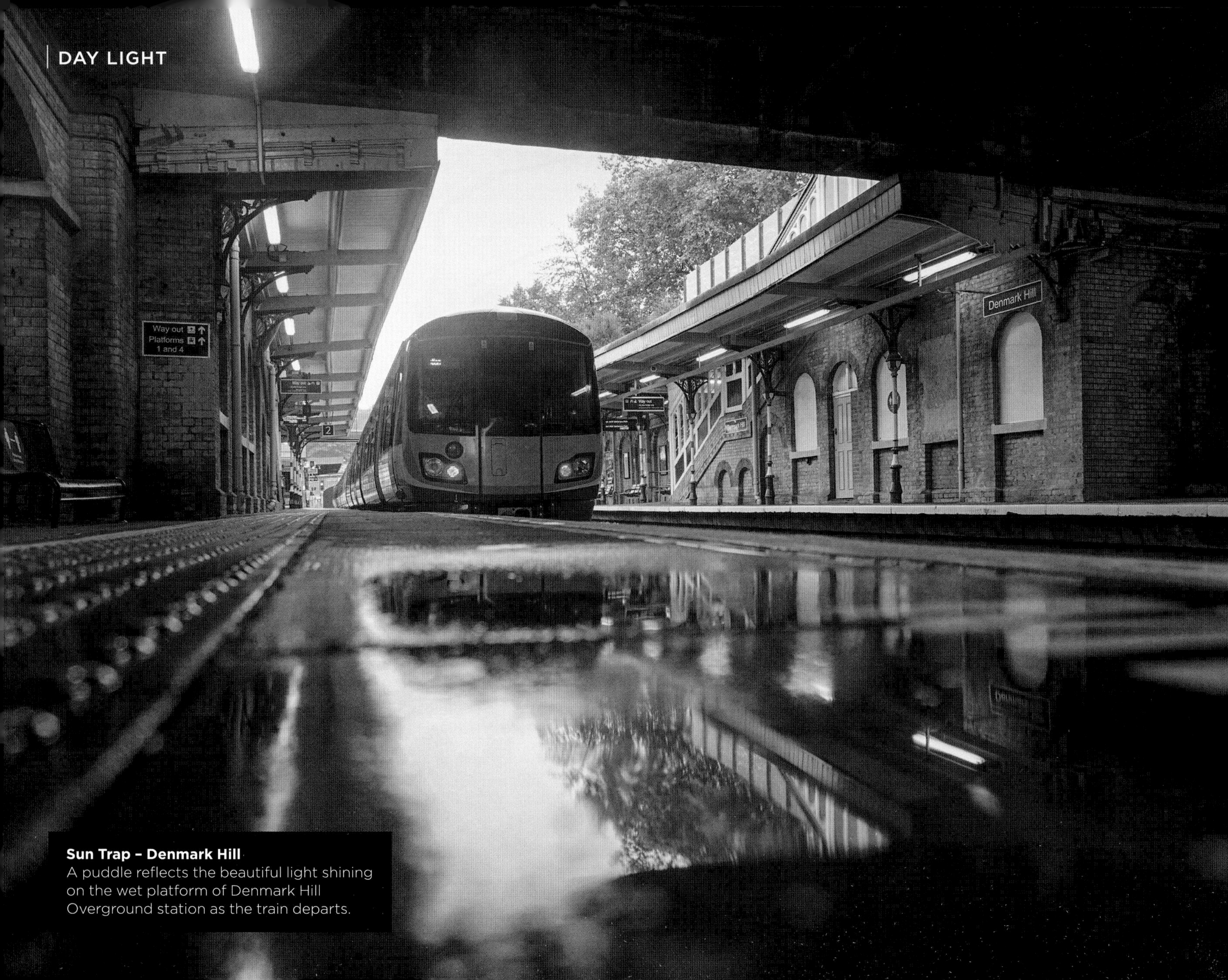

Sun Trap – Denmark Hill
A puddle reflects the beautiful light shining on the wet platform of Denmark Hill Overground station as the train departs.

mist-filtered light
slowly the space fills
with electric ice

John Hawkhead

Approaching the Hill – Harrow-on-the-Hill
Harrow-on-the-Hill station is especially pleasing to look at during the evening. Here's a puddle reflection photo of the surprisingly handsome exterior, which was designed by Stanley Heaps.

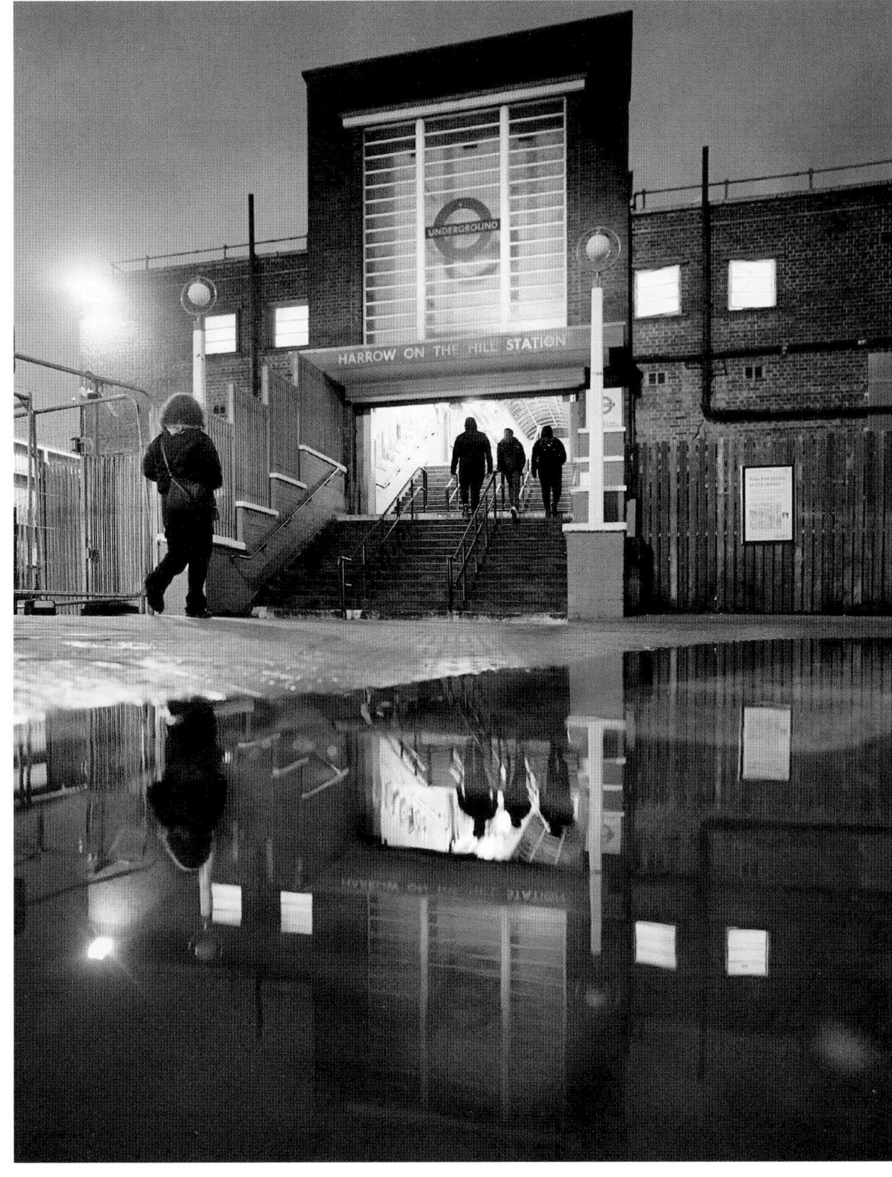

UNDERGROUND
HARROW ON THE HILL STATION

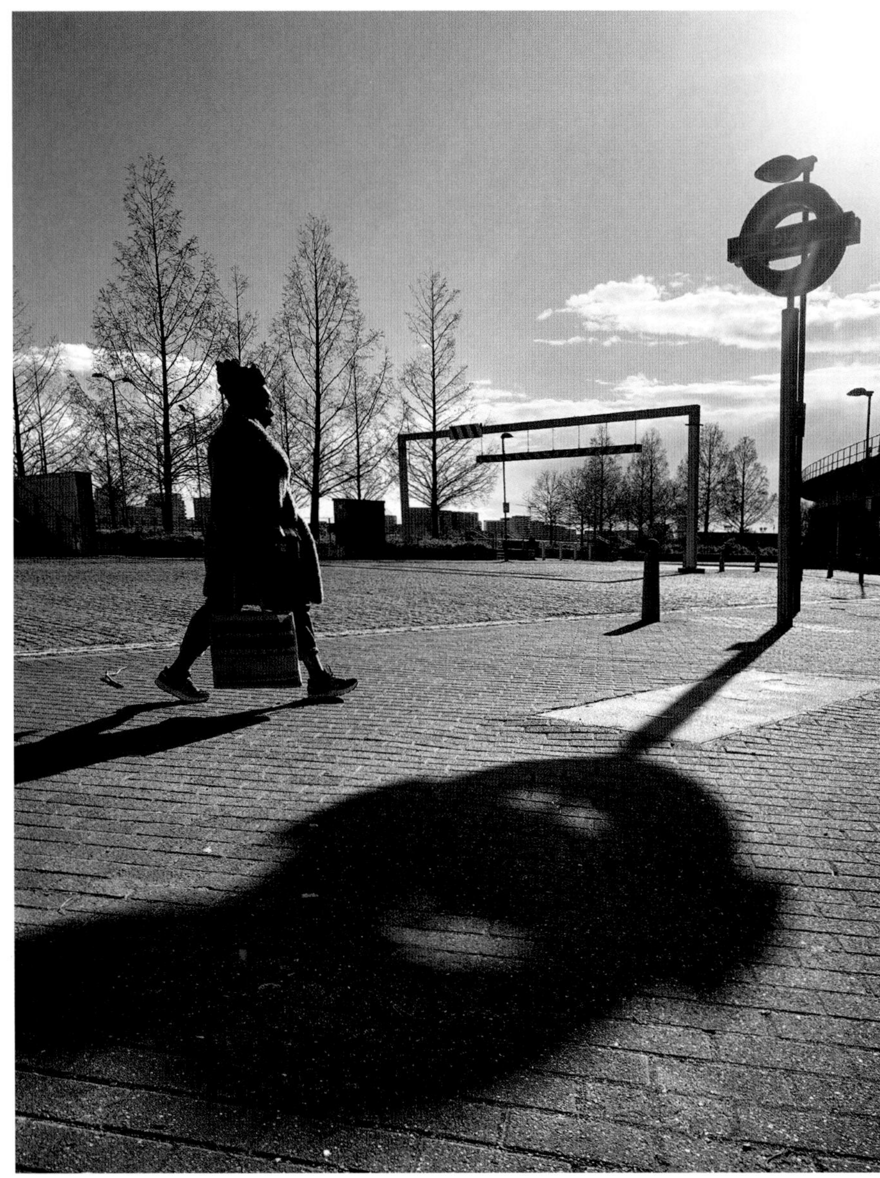

Shadows – Royal Albert DLR
Depending on the sun's position, you can photograph exciting compositions in the most surprising locations. Here, the low afternoon sun crafted the long, elegant shadows of the station's roundel.

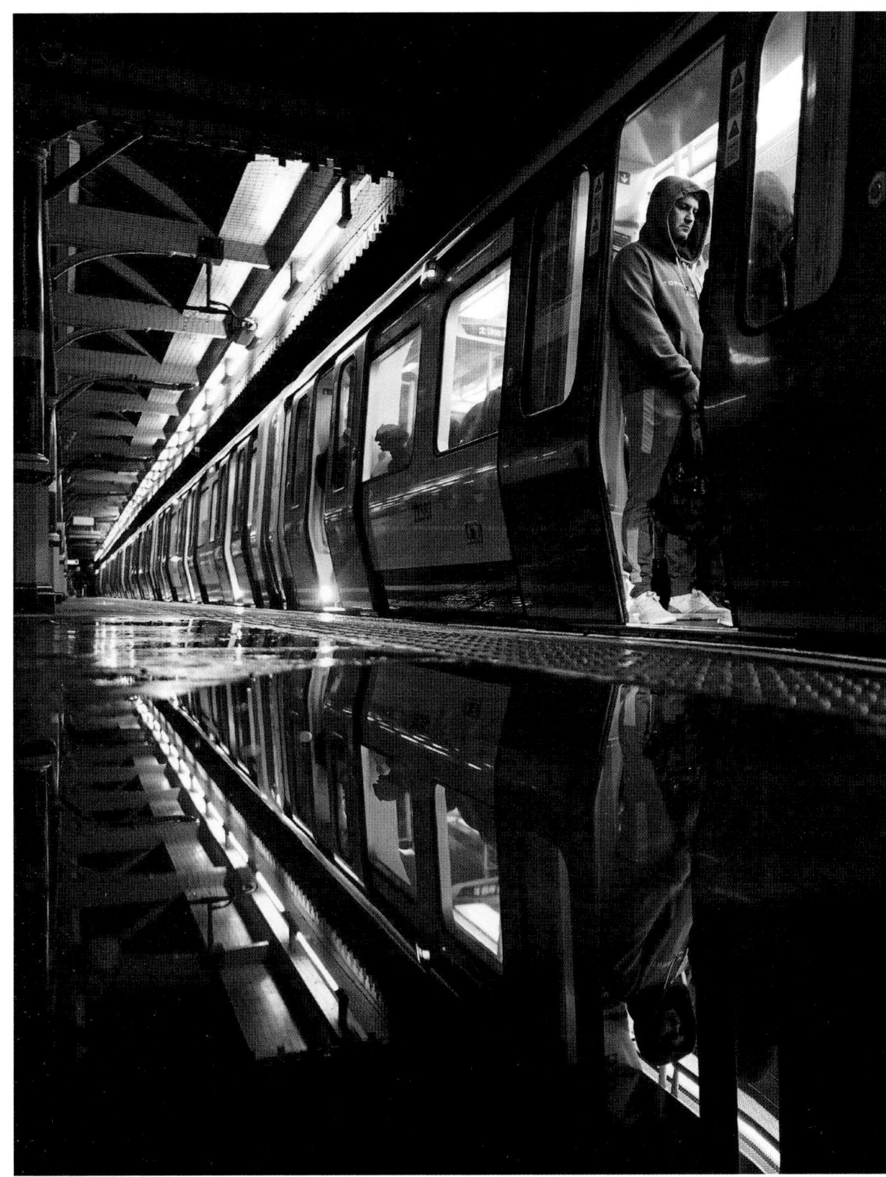

Reflecting – South Kensington
This night shot was taken as the train doors closed at South Kensington station. The wet platform reflected the wonderful lights in the canopy and the interior of the train.

Shadow Lines – Cyprus DLR
I purposely visited this location to capture the beautiful light at Cyprus DLR station and appreciate the dramatic shadows created by the architecture.

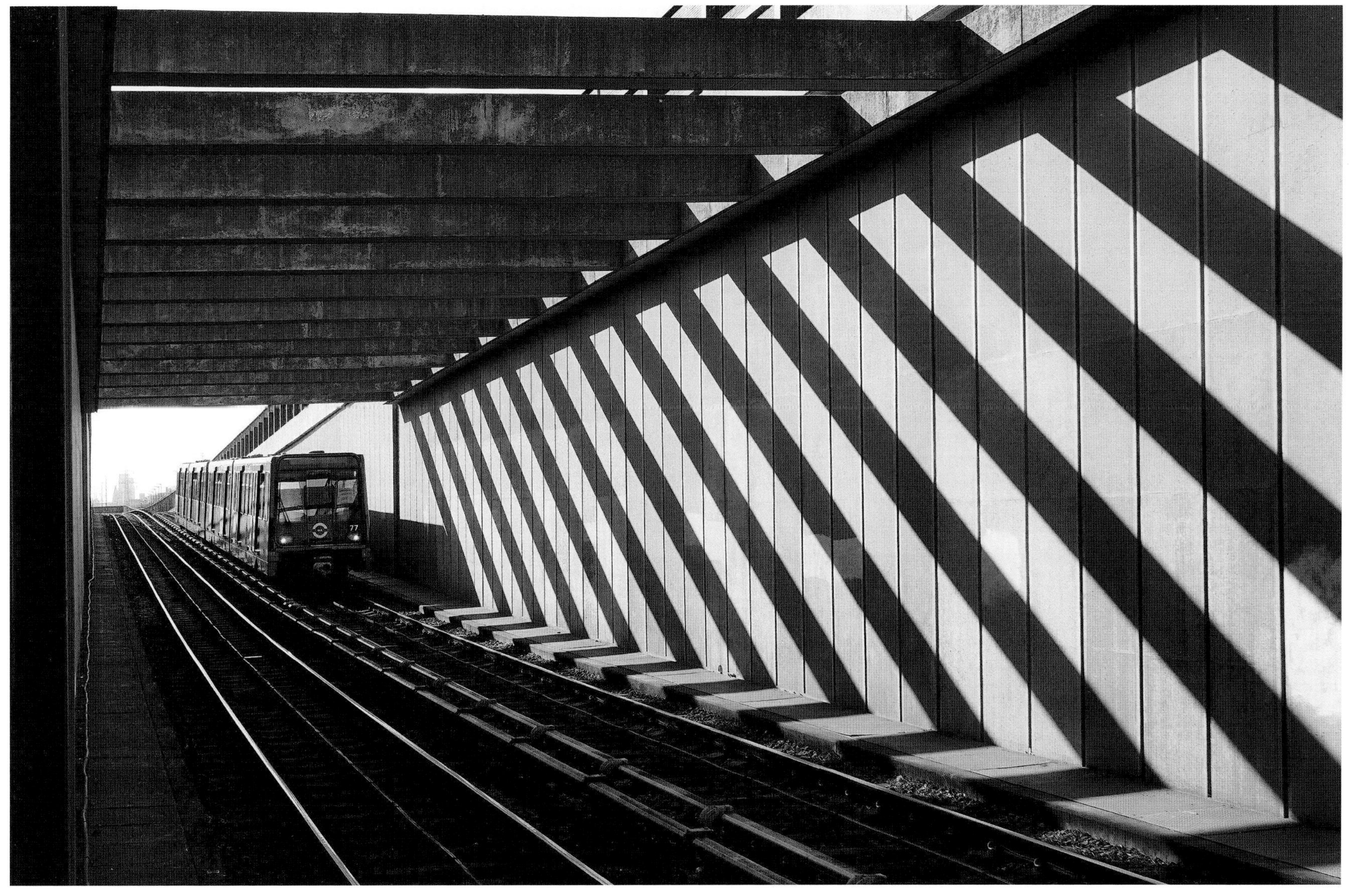

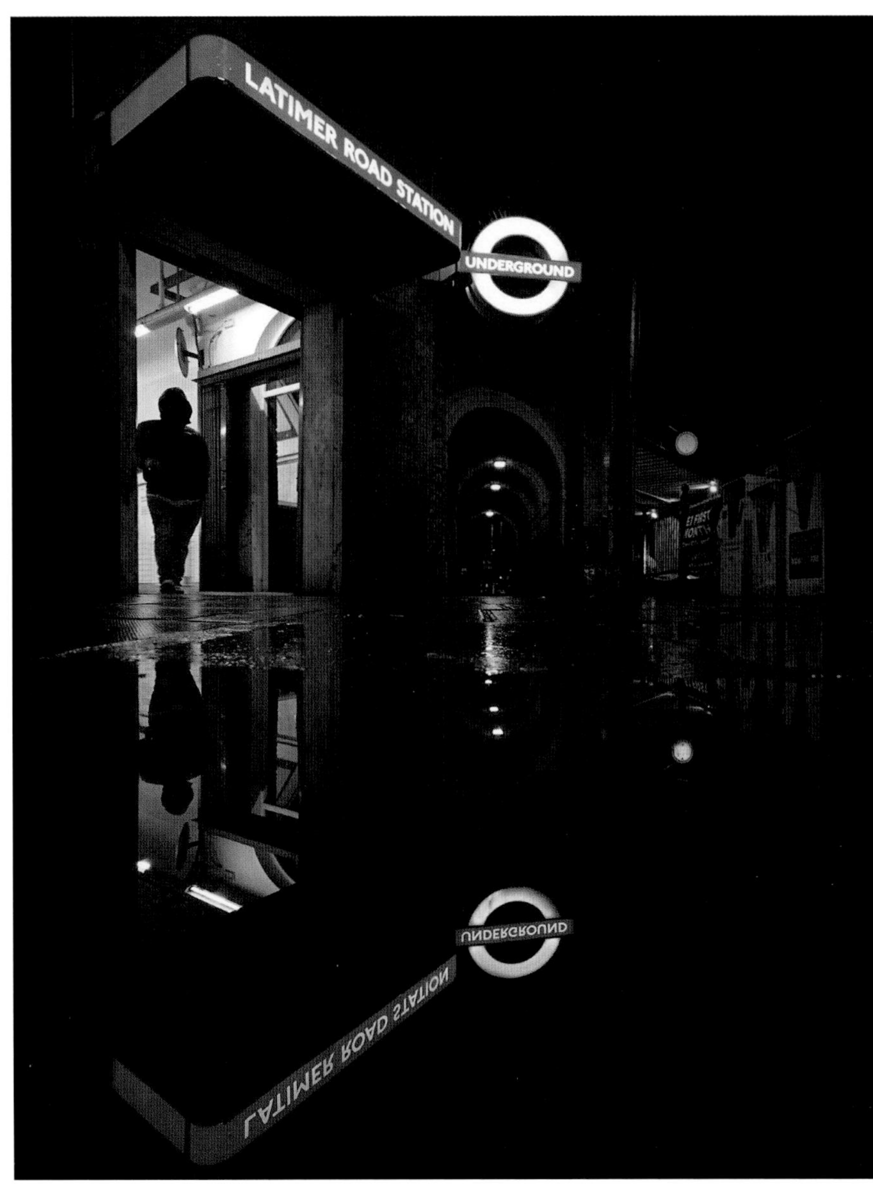

Late Night – Latimer Road
A puddle reflection of the entrance to Latimer Road station.
I love the shapes created by the mirrored architectural lights
and the mood created by the silhouetted hooded man.

rolling in
on the midnight train
wheels of moonlight

John Hawkhead

Arch – Crystal Palace
Looking down the stairs at Crystal Palace Overground station, the late afternoon sun creates the long shadows that stretch across the platform.

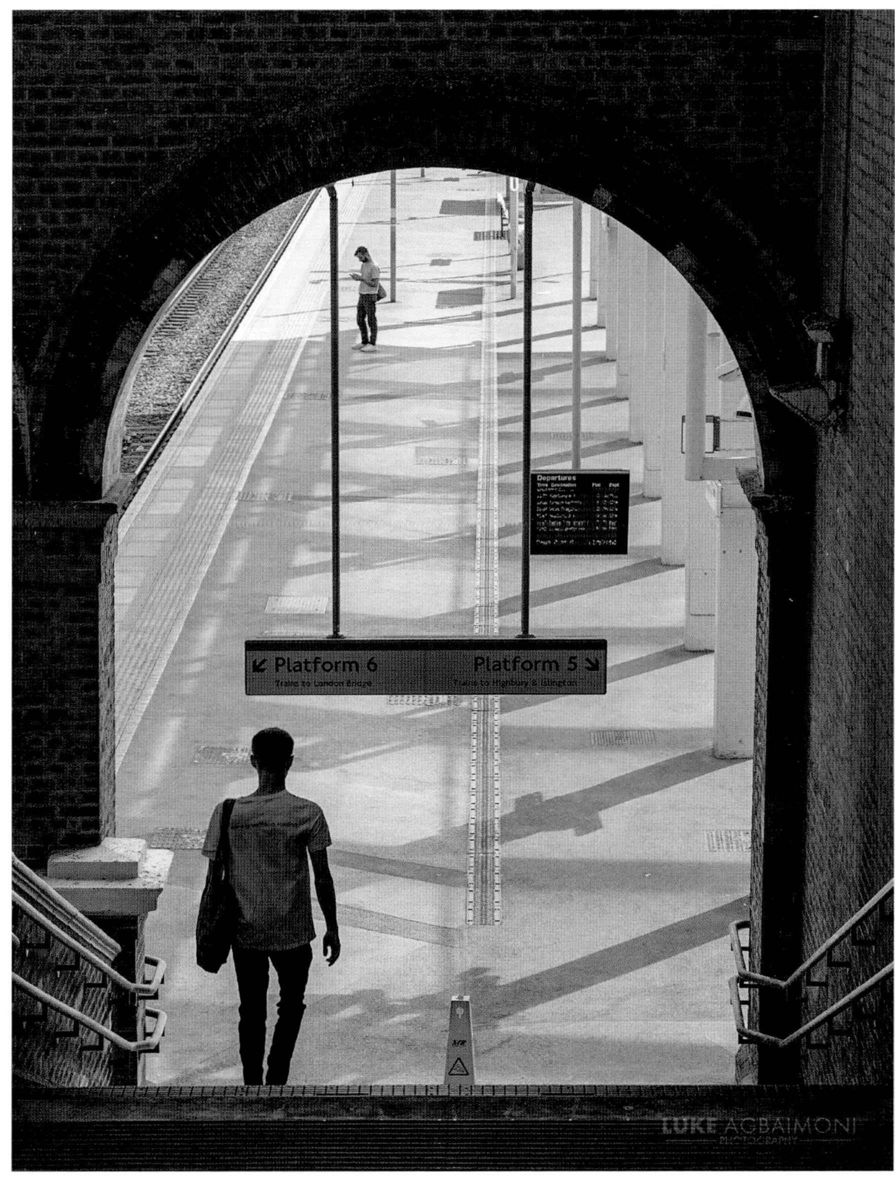
Departures
Platform 6
Trains to London Bridge
Platform 5
Trains to Highbury & Islington
LUKE AGBAIMONI
PHOTOGRAPHY

Bike Reflection – Shadwell Overground
Due to the bicycle lane in front of Shadwell station, many bikes pass by. The large number
of bicycles gave me plenty of opportunities to nail the timing of this puddle reflection shot.

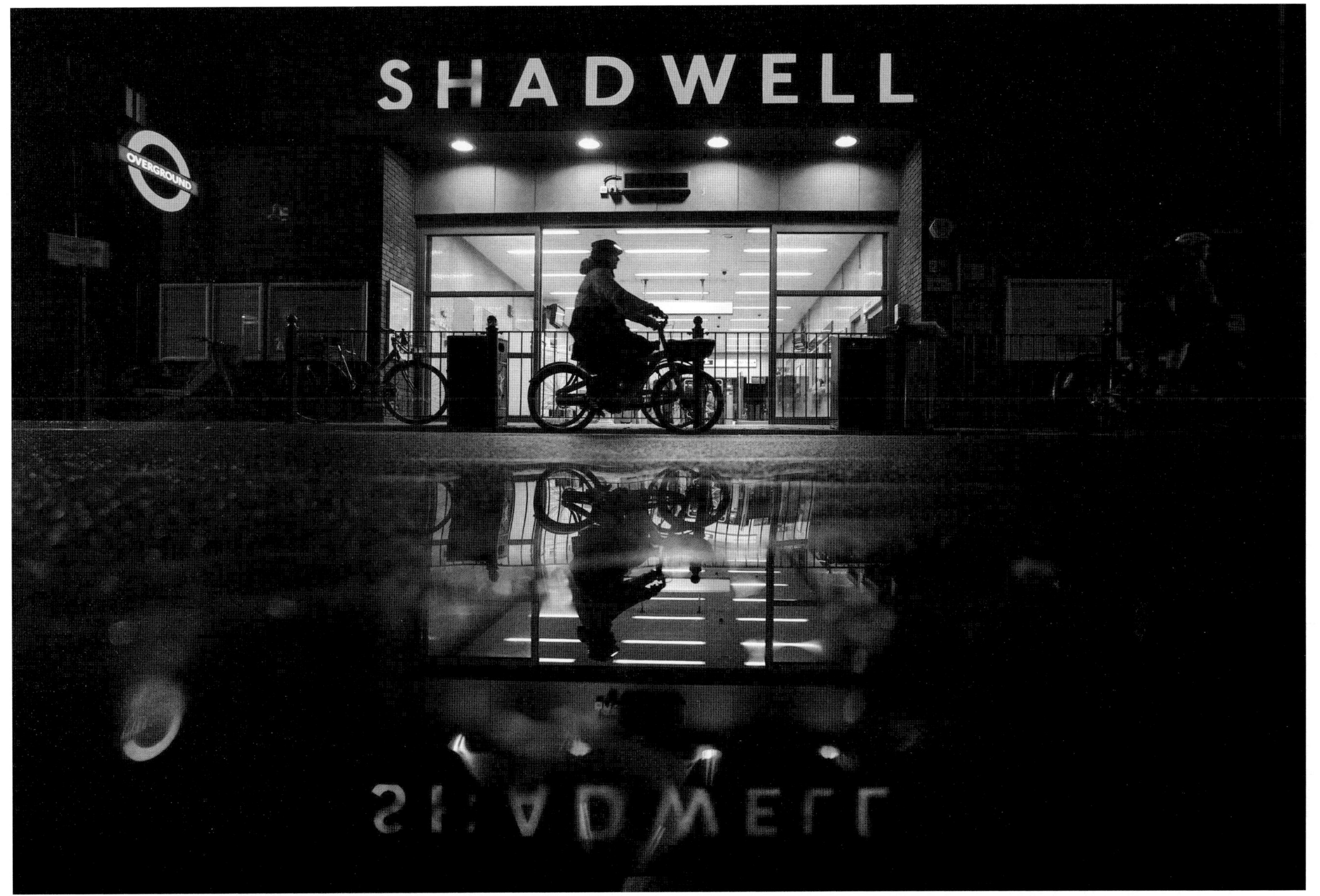
SHADWELL
OVERGROUND

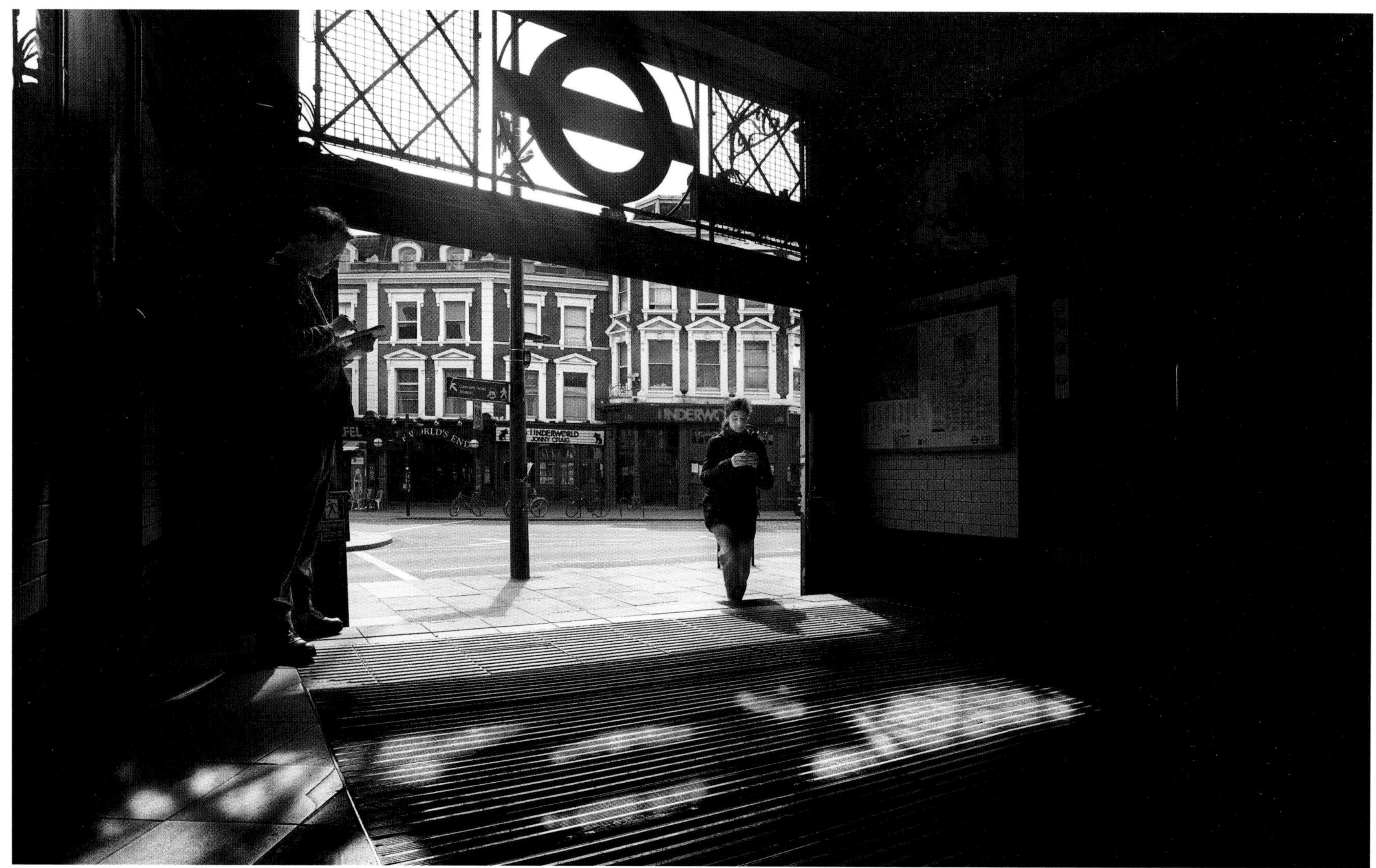

Roundel Shadow – Camden Town
I've always liked the roundel design of the entrance to Camden Town station, but I've only recently discovered that it casts pleasing shadows on the floor if you're there at the right time.

4

STRAIGHT LINES/CURVES

The Tube map is a complex maze of intricate intersecting lines and curves, with each train line guiding you to locations that have unique identities and historical relevance. The architectural structures of the stations you'll encounter on your journeys can vary quite dramatically depending on where you are on the network. I love the contrast between the sharp angular lines of long square pathways and the sweeping curves of spiral staircases and round tunnels.

Here's a collection of images exploring straight lines and curves.

Photo Hints

Straight Lines and *Curves*: The angle and plane of your camera can dramatically affect the overall composition of the image when crafting shots. Explore different positions and see how they change the feel of the photo.

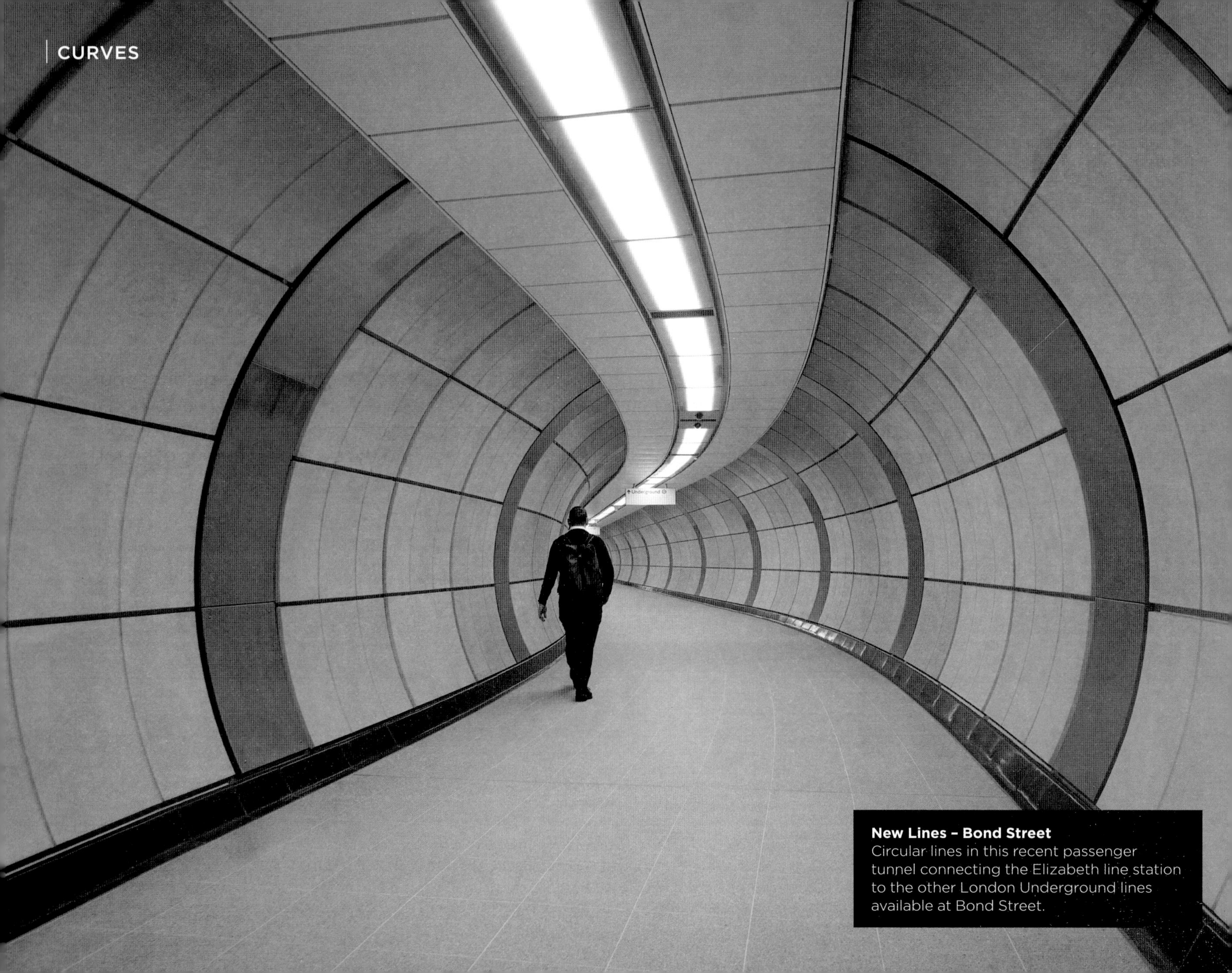

New Lines – Bond Street
Circular lines in this recent passenger tunnel connecting the Elizabeth line station to the other London Underground lines available at Bond Street.

how it all started
the curve of a railway line
joining another

John Hawkhead

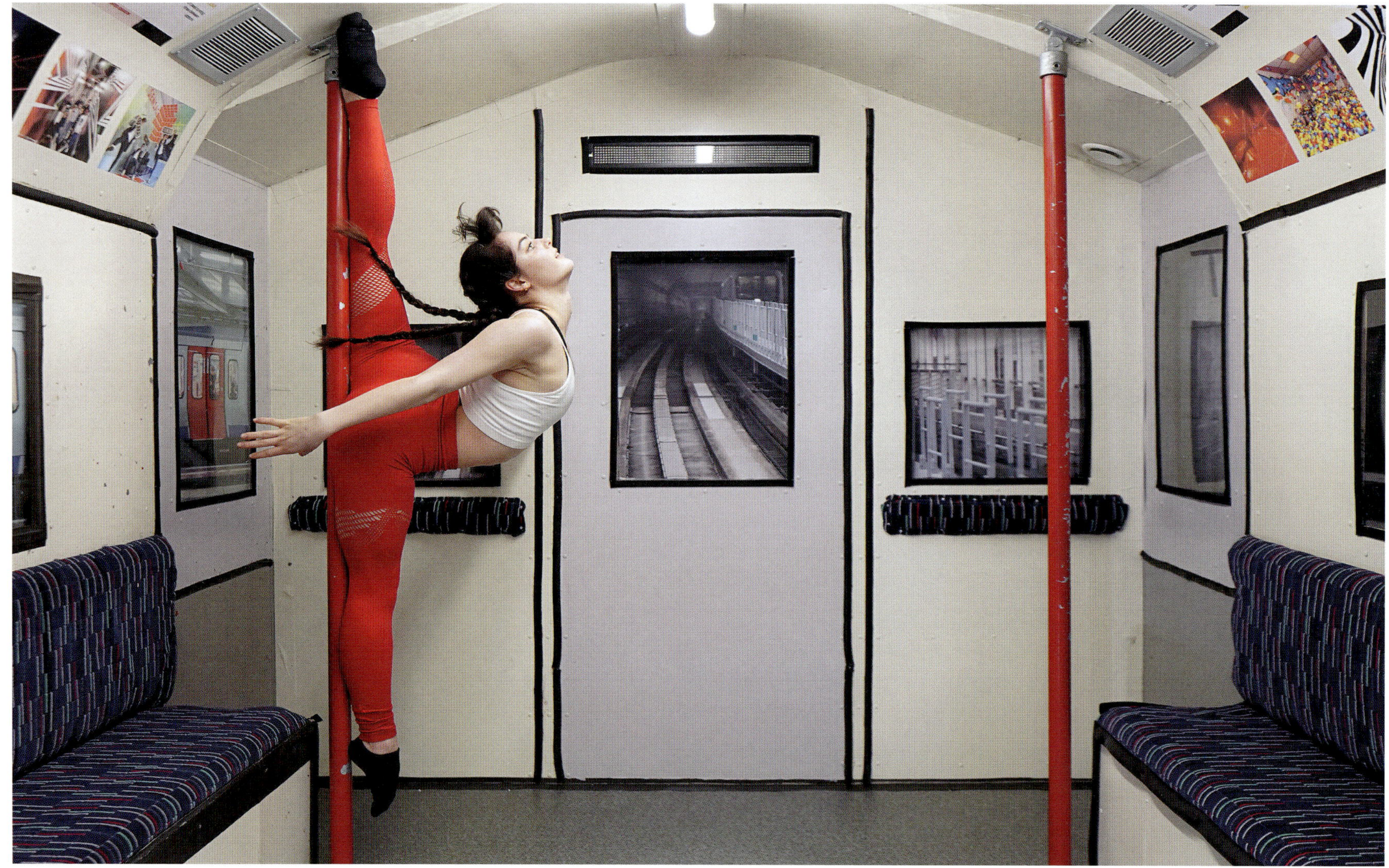

Pole Position – Vauxhall

Anna complements the vertical lines in what looks like a train carriage but is surprisingly a sideways escape room at the Sensas Sensory Labs near Vauxhall and Nine Elms stations.

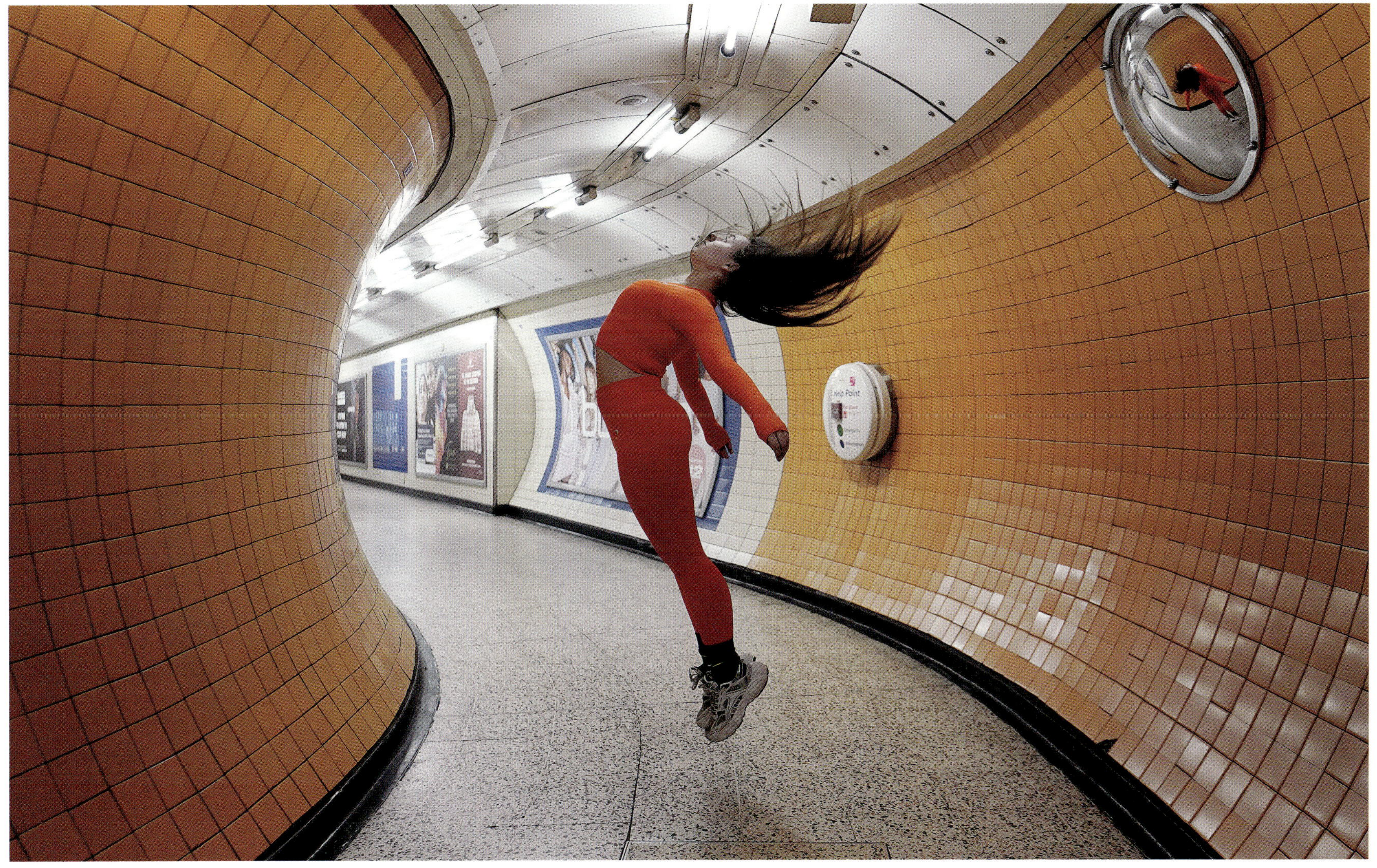

Orange – Bond Street
There is a surprising splash of orange in this curvy passageway at Bond Street station, where Andreya interacts with vibrant tiles.

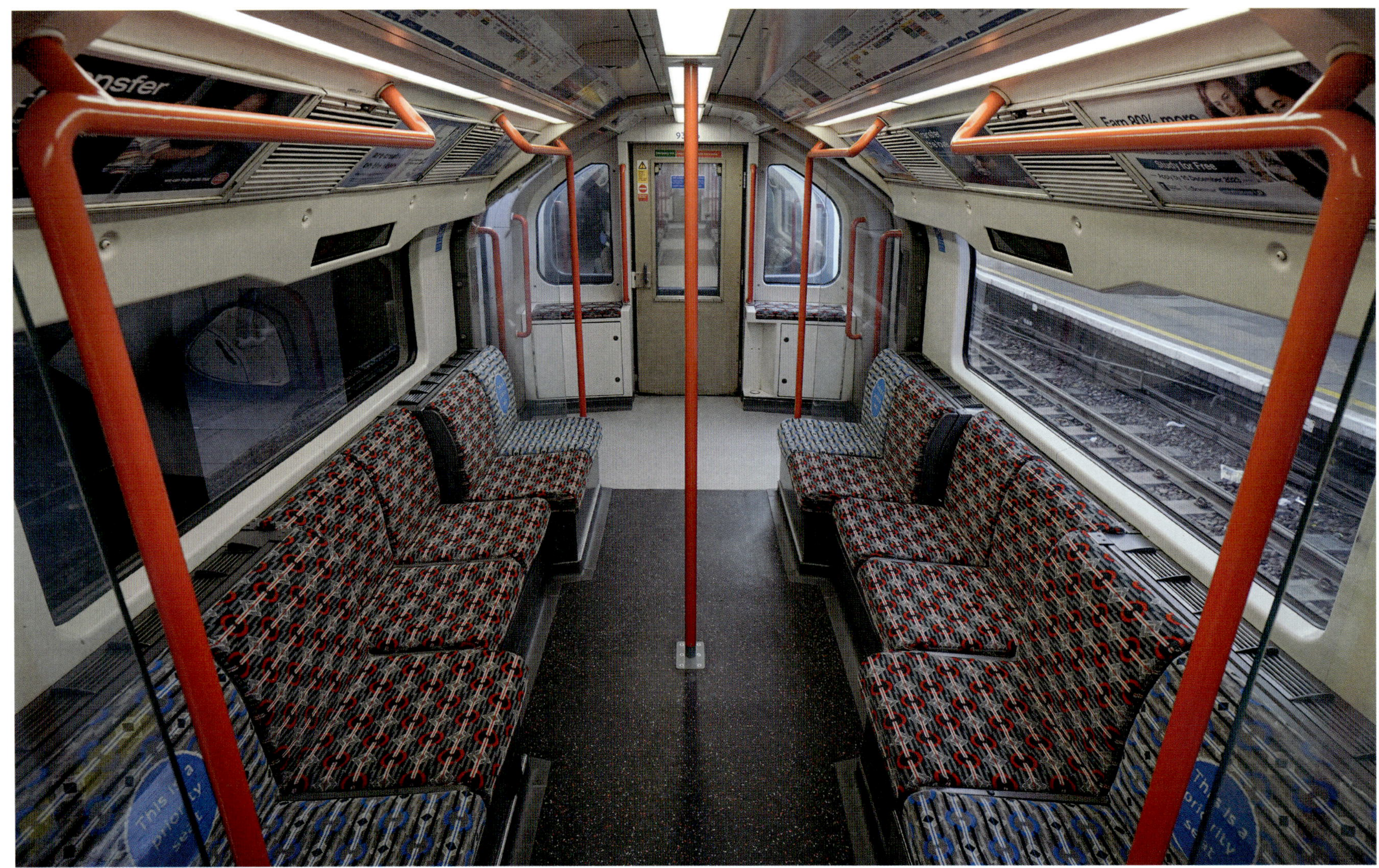

Central Line – Hainault
The recent refurbishment of the Central line train carriages includes a handsome new moquette design.

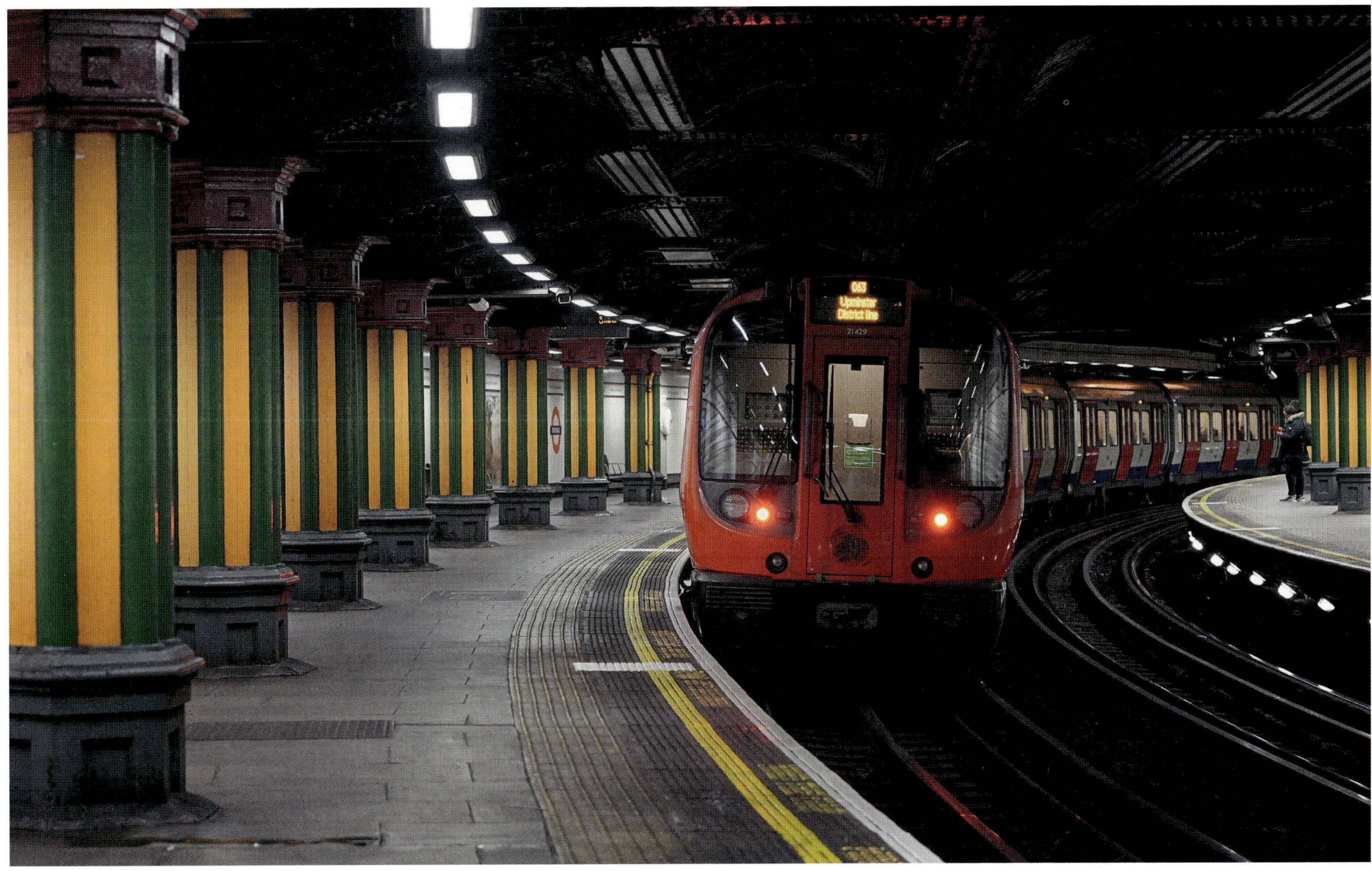

Curves – Bow Road
A train departs the curved platform at Bow Road station. This Grade II listed building opened in 1902 and has various striking features, including these colourful hexagonal pillars.

Between Trams – Beckenham Junction
You may encounter two trams parked at Beckenham Junction tram stop if you're lucky –
they don't naturally line up, so you must snap one in motion to capture them in symmetrical alignment.

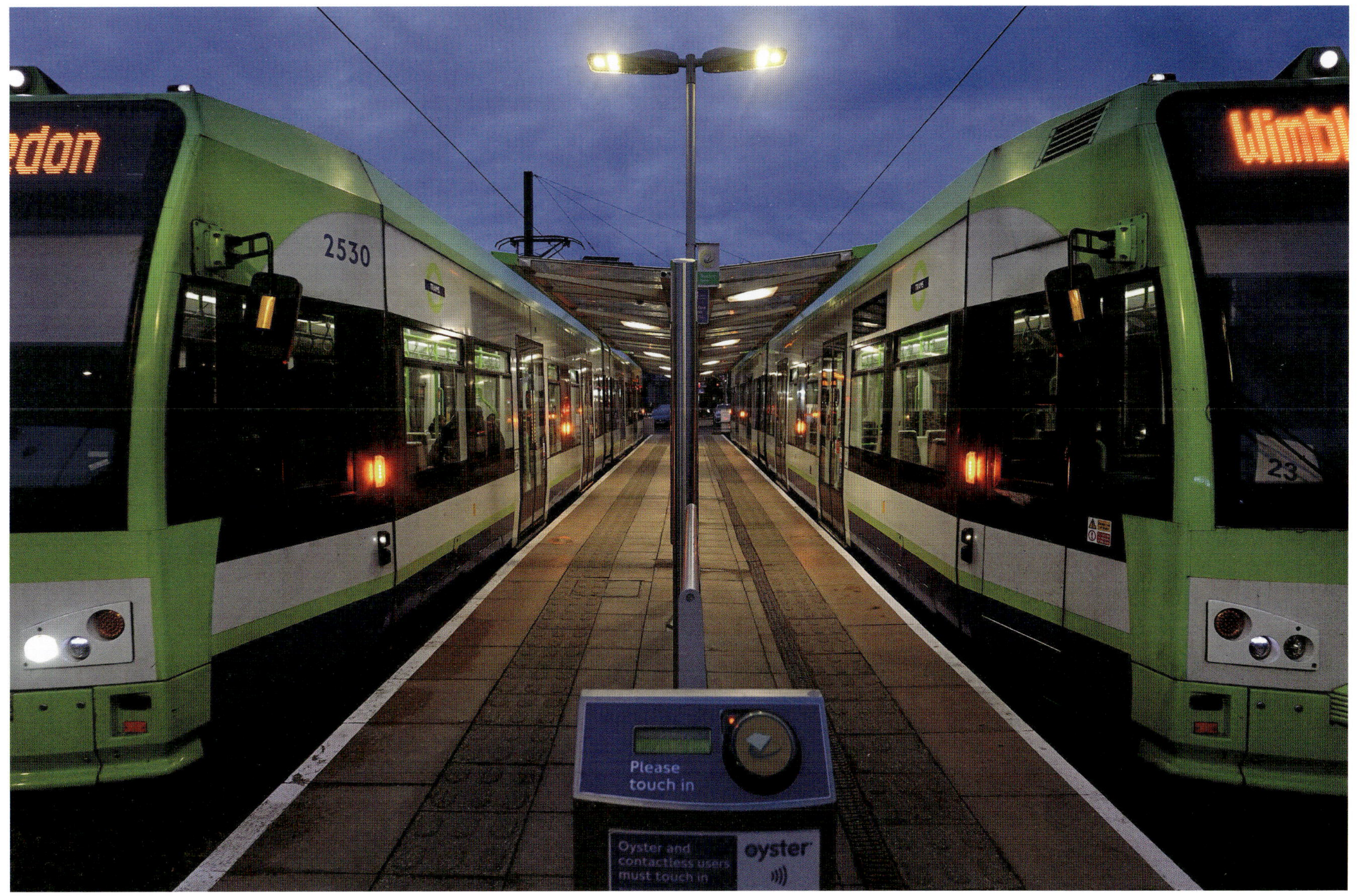

edon
2530
Wimbl
Please
touch in
Oyster and
contactless users
must touch in
oyster

Red Arch – Bank

Gymnast Anna poses beneath the red Greathead tunnelling shield at Bank station, used during the construction of the original Waterloo & City line in 1898. I love how this easy-to-miss historic structure remains on display like a bonus museum piece.

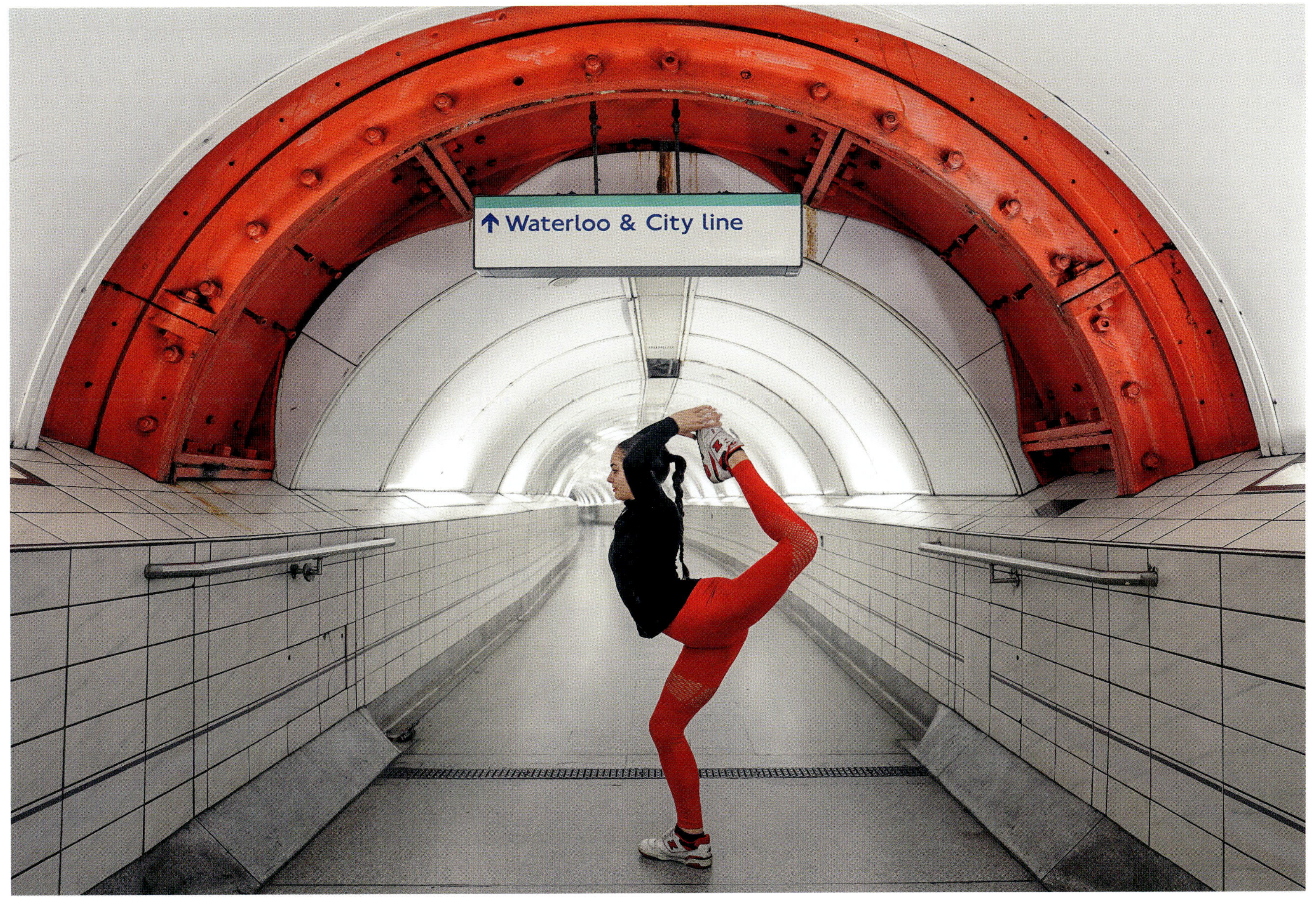
↑ Waterloo & City line

Mirrored Building – Waterloo
An abstract observation as I looked up towards the London Underground roundel outside Waterloo station.
The reflections on the glass highlight the dynamic lines of the buildings opposite.

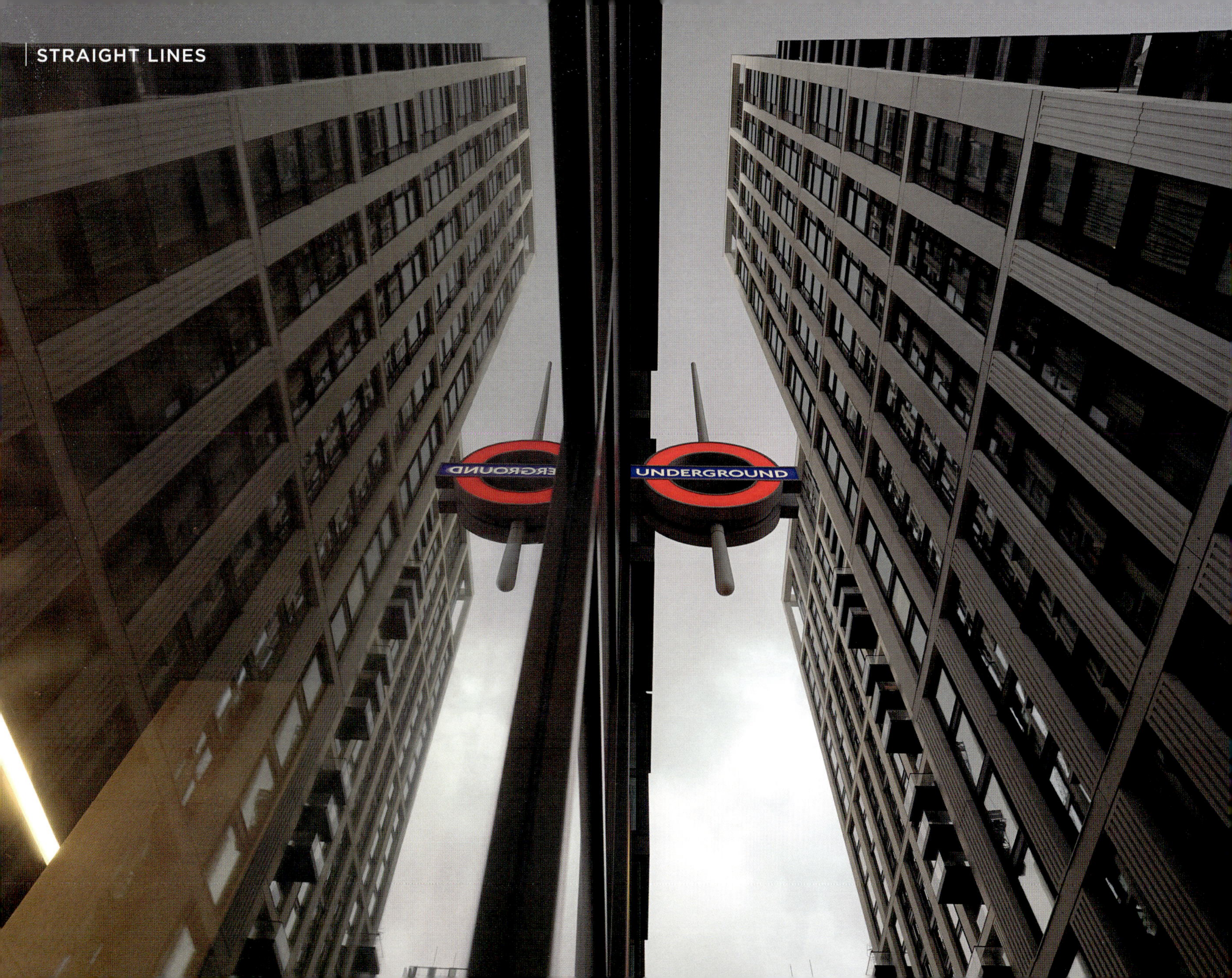
UNDERGROUND

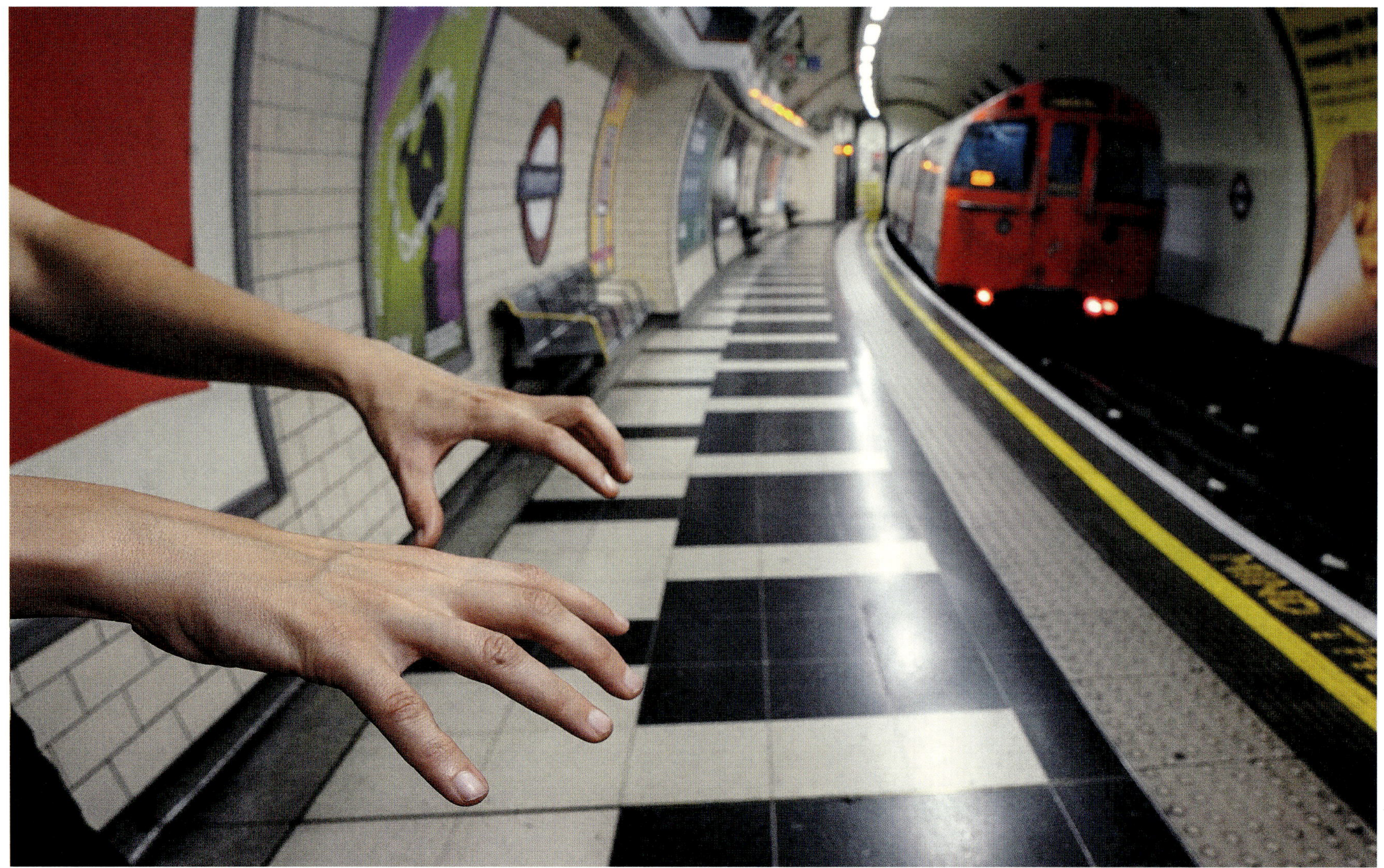

Playing the Platform Piano – Waterloo
This is a fun photo of musician Fiona Fey interacting with the piano-key floor tiles on the curved Bakerloo line platform at Waterloo station. This easy-to-miss feature was installed between 1986 and 1988, when the station was retiled based on Christopher Tipping's 'theatrical themes' design.

along the old track
shimmerings of frost
a slow curve of stars

John Hawkhead

Spider – Oxford Circus
The lines and angles of this staircase at Oxford Circus station resemble the shape of a giant spider.

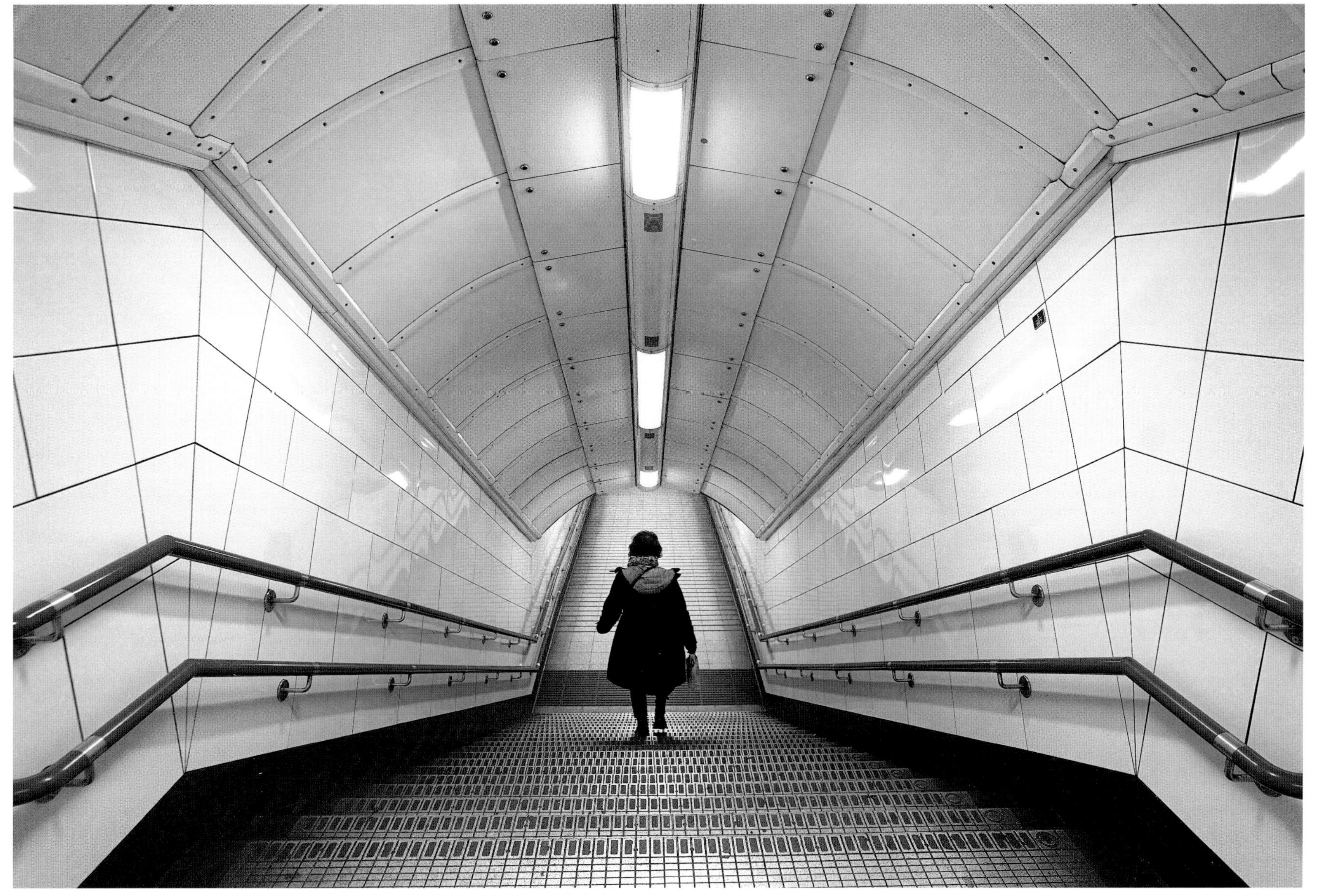

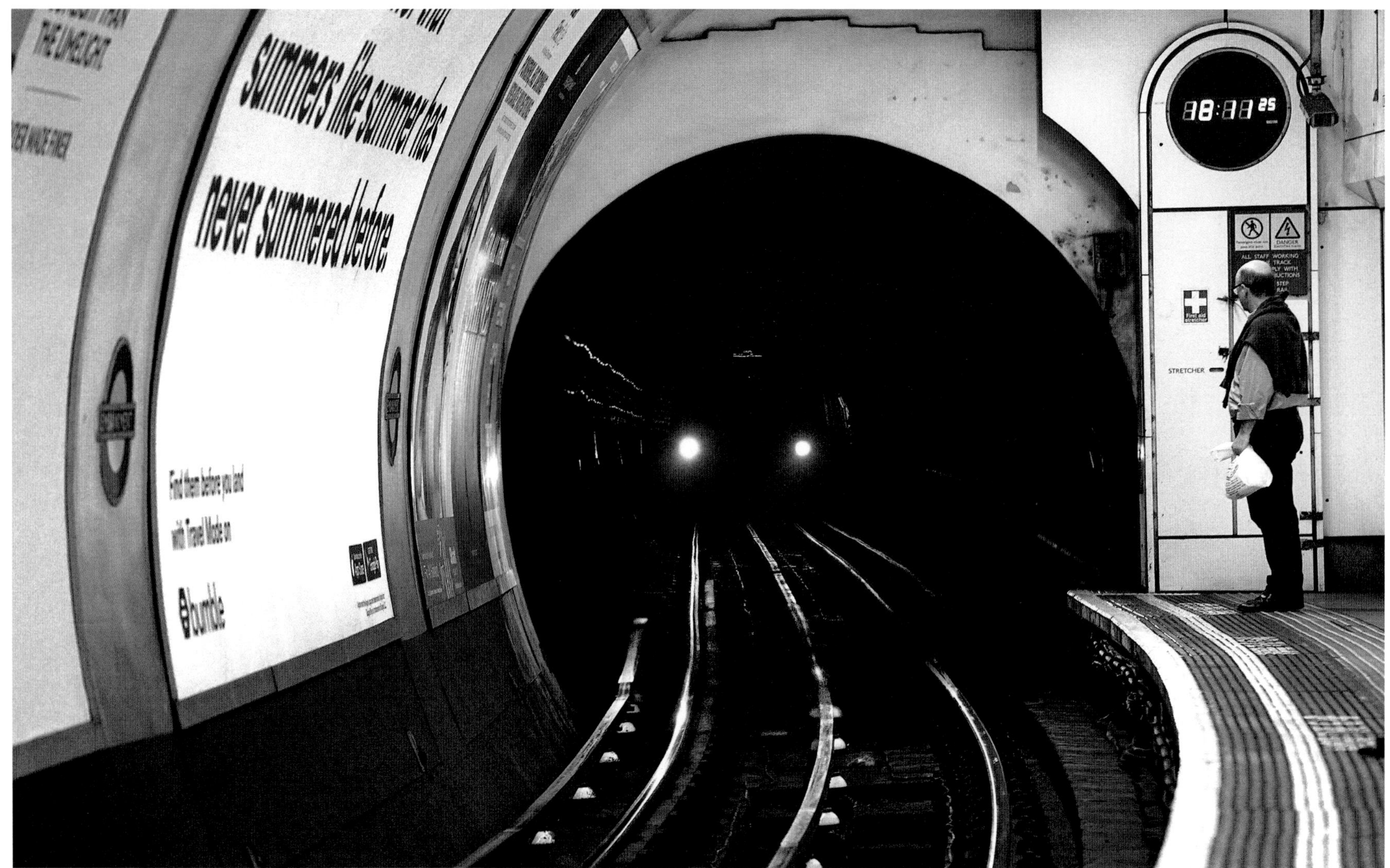

Waiting – Embankment
I love the anticipation of a train's arrival: the gust of wind and distant noises, the dark track being painted by the eager lights of the incoming train. I like the clock on this platform and how its shape repeats throughout this composition.

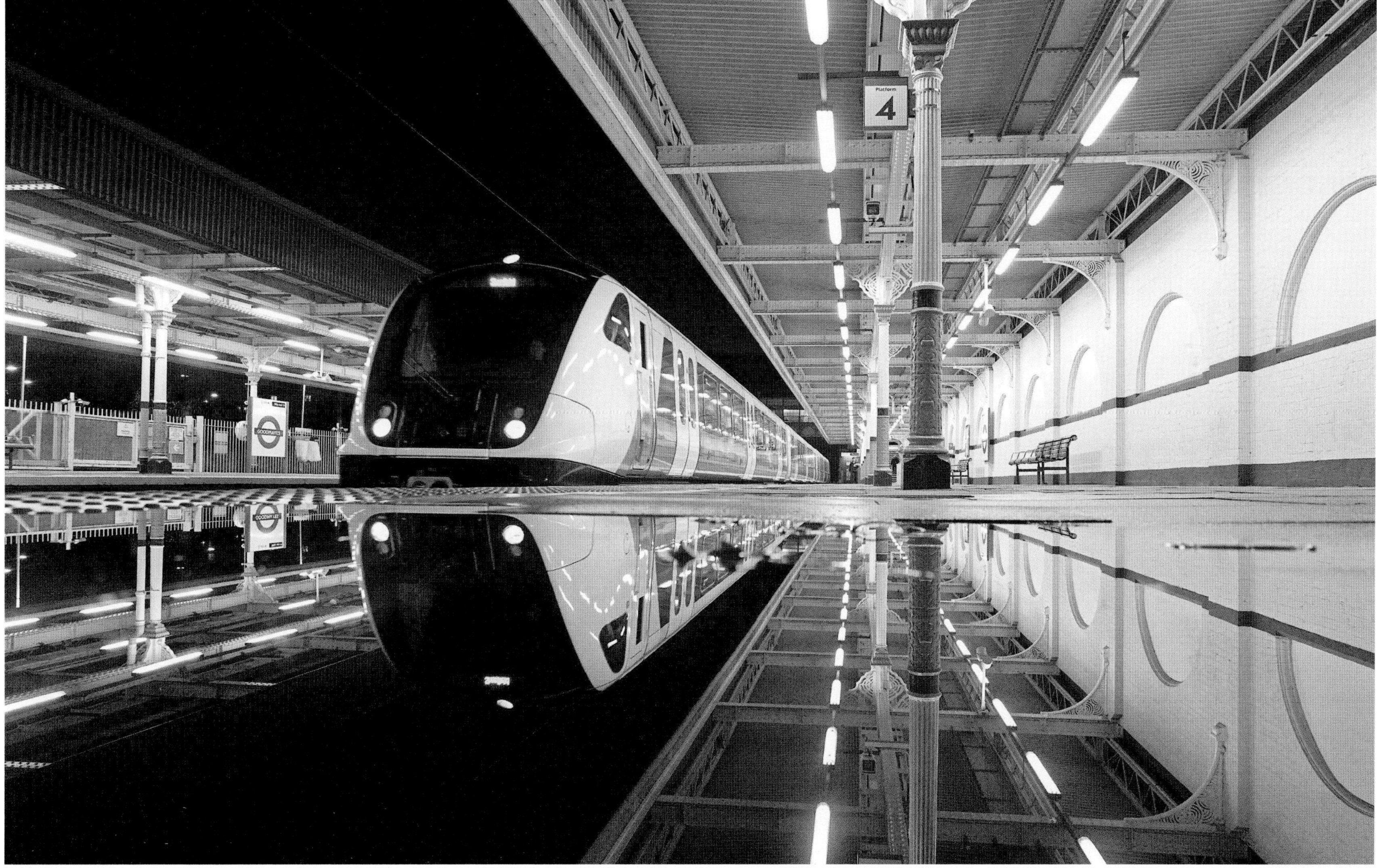

Reflections – Goodmayes
I couldn't resist capturing the reflections in this puddle at Goodmayes; they highlight the incredible architectural lines of this Elizabeth line station.

Curves & Lines – Canada Water

I always enjoy watching the partial view of the circular roof slowly reveal itself as you head up the escalators.

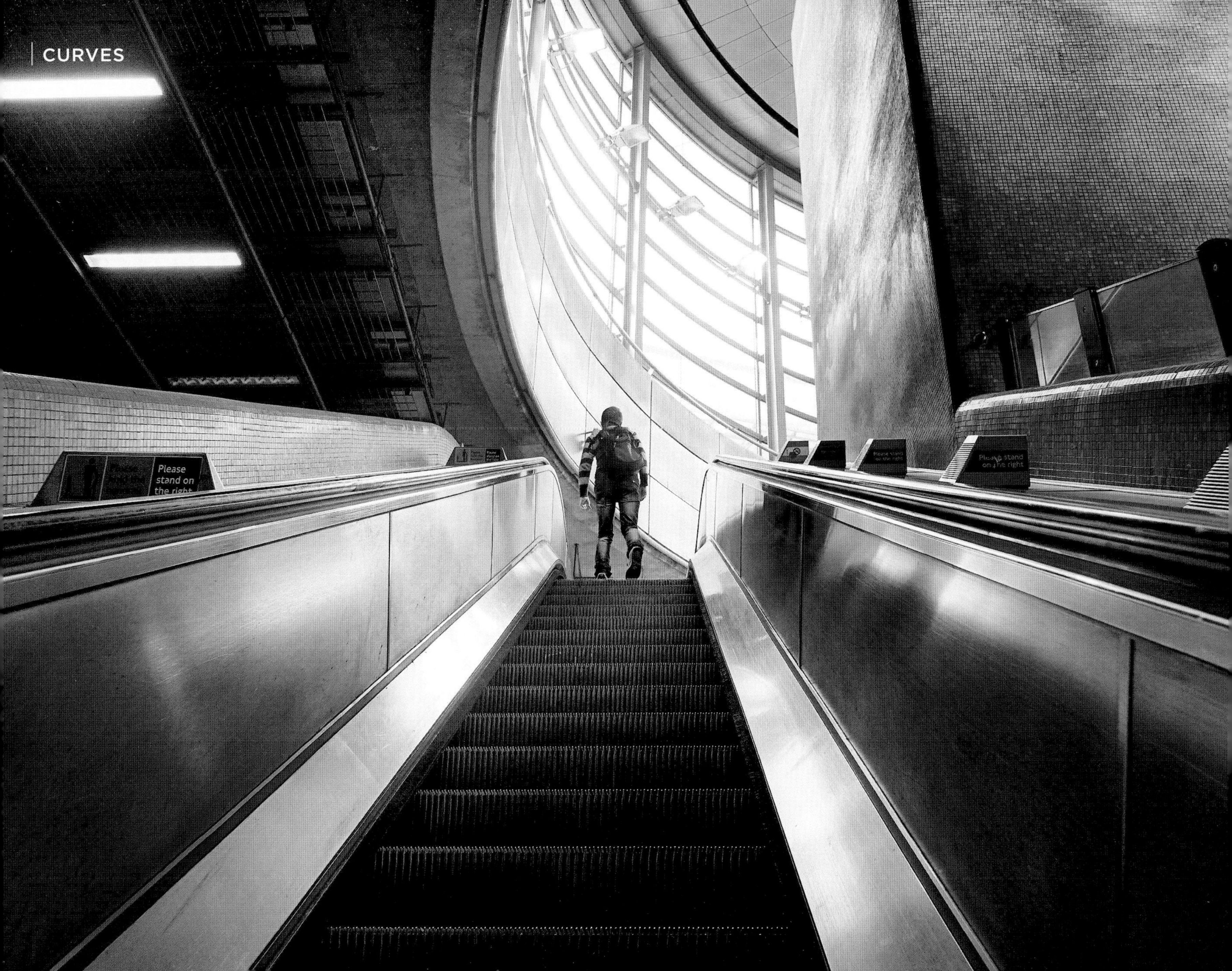
Please
stand on
the right
Please stand
on the right
Please stand
on the right

Stairs – Moorgate
The powerful architectural lines on this staircase lead downwards from the Northern line platform to Liverpool Street Elizabeth line station.

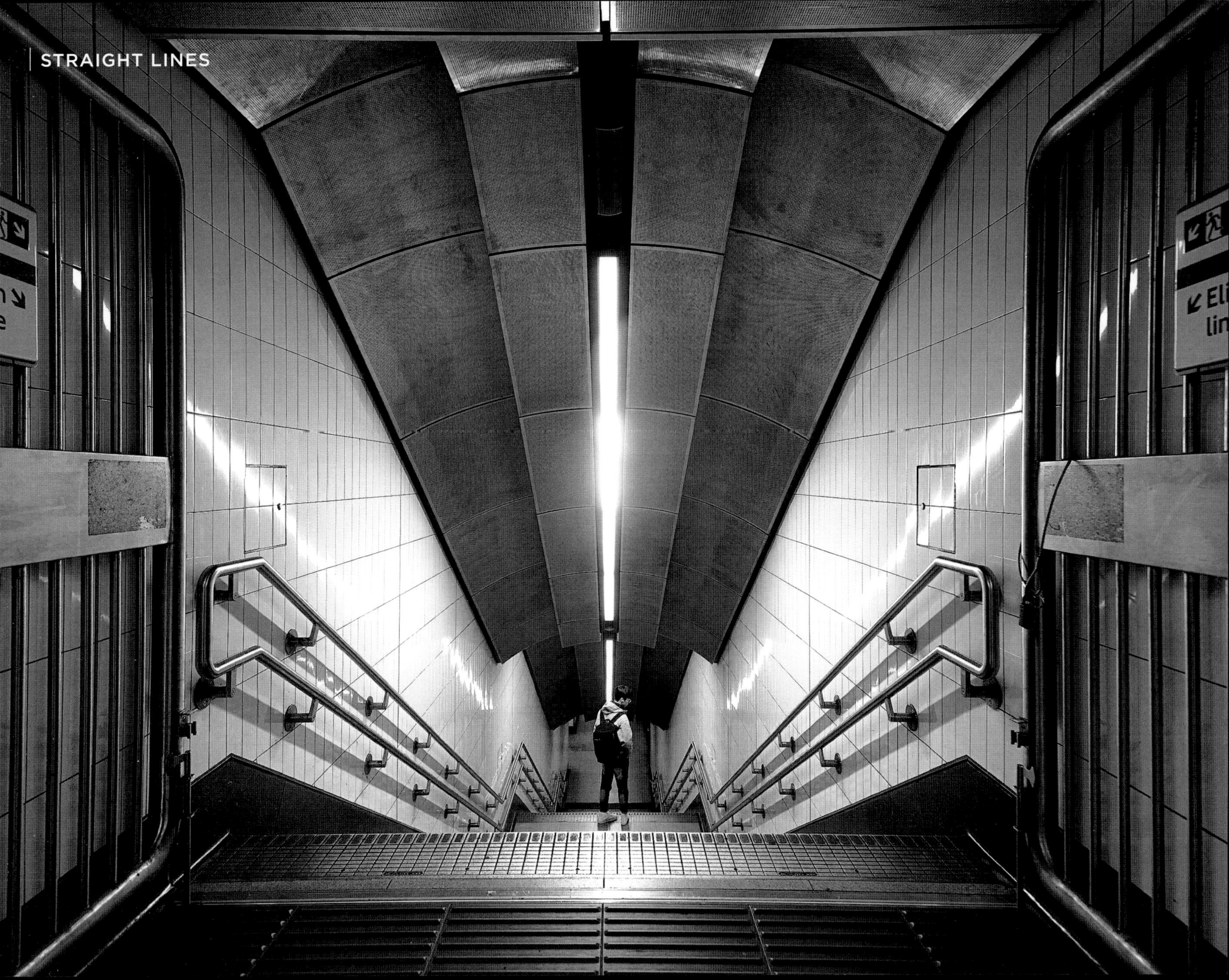
STRAIGHT LINES

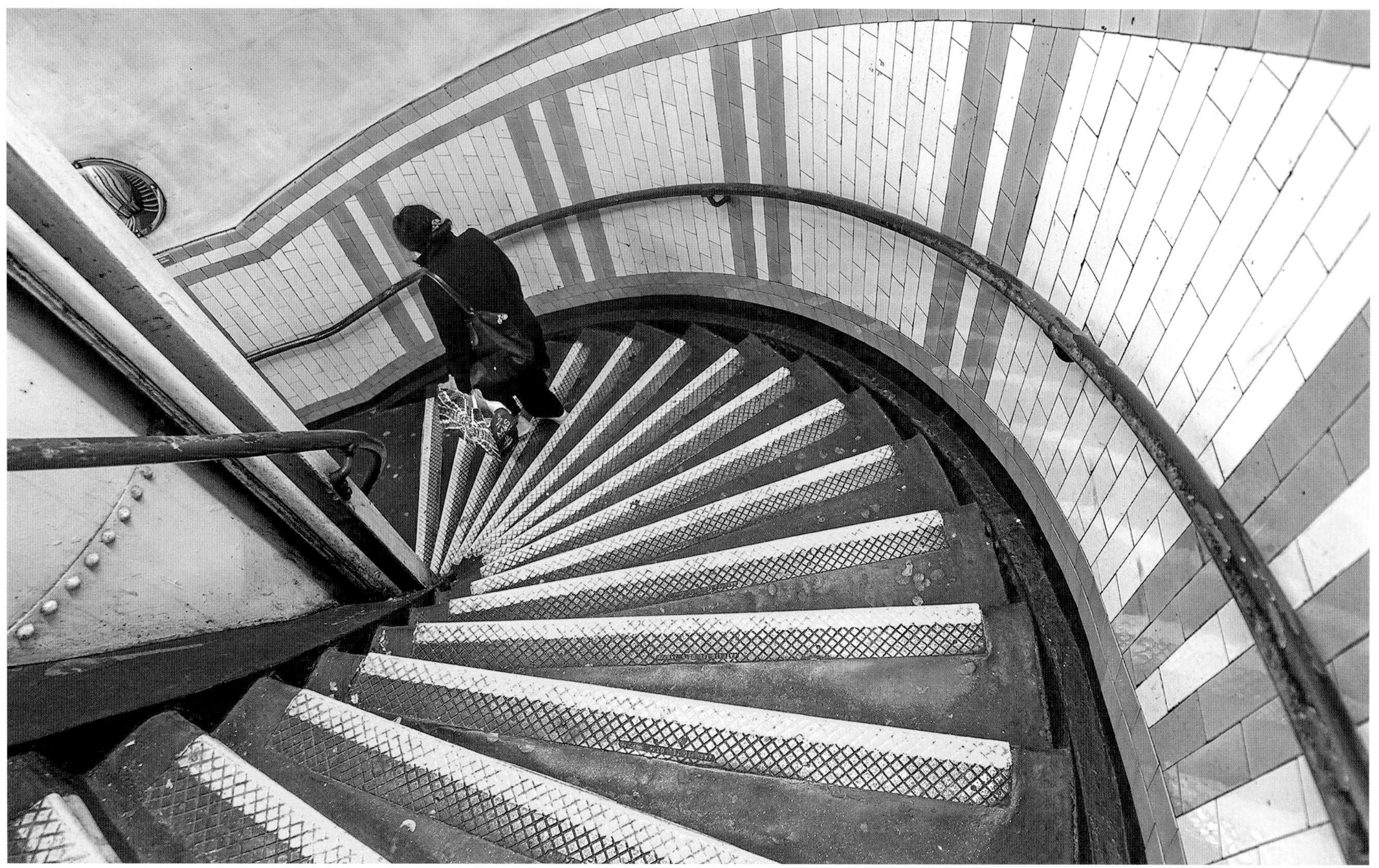

Spiralling Underground – Camden Town
These beautiful spiral staircases can usually be found at fire exits, but are also occasionally hidden elsewhere in the station.
I love these original tiles at Camden Town's fire exit.

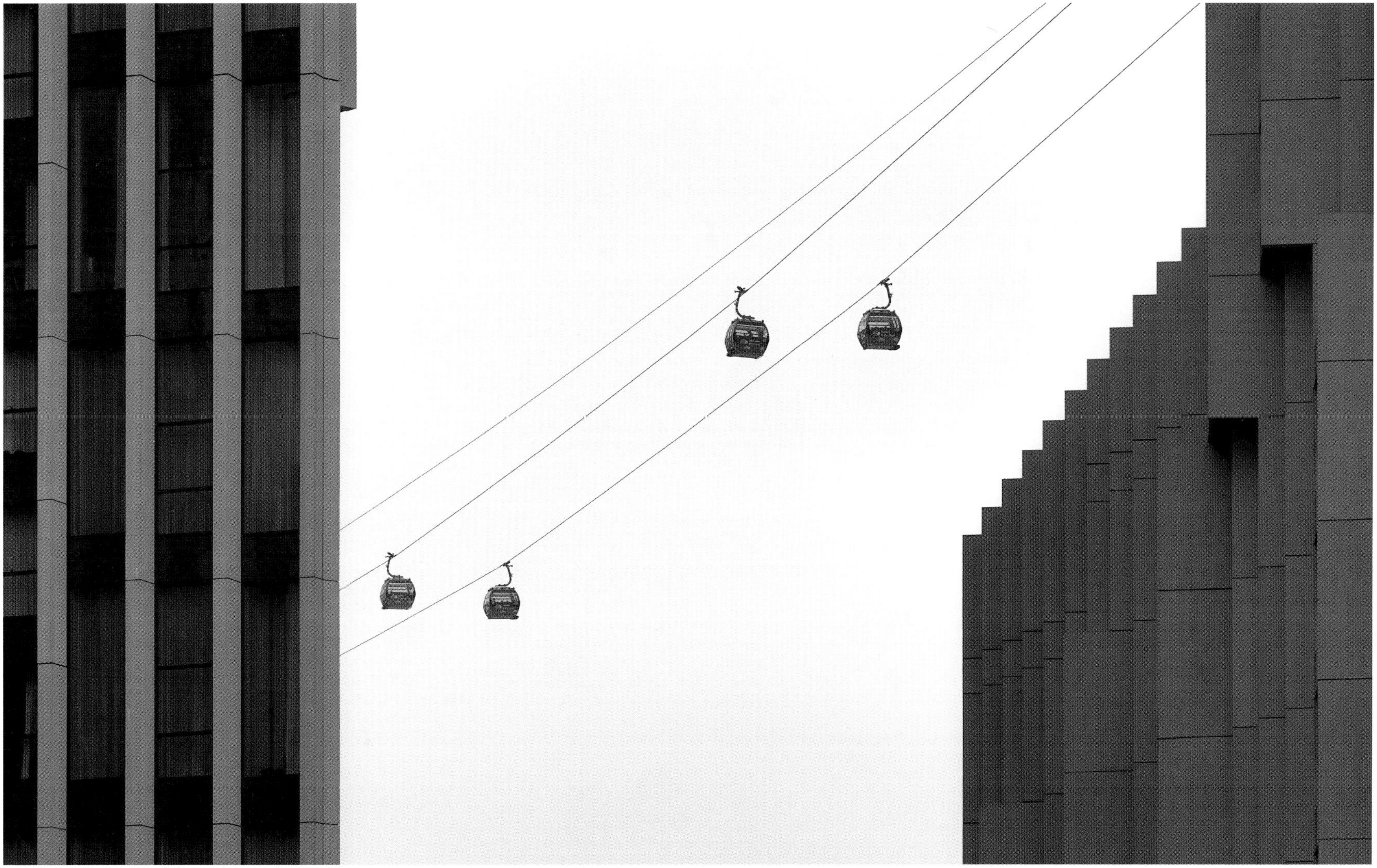

Cable Lines – Greenwich Peninsula
This minimalist photo was taken from the Greenwich Peninsula, looking between two buildings at the London cable cars near North Greenwich station. The cable cars are a surprising addition to the city's skyline.

Curves – Mornington Crescent
A photograph capturing the beautiful curves in the ceiling of this tunnel located at Mornington Crescent station. I also enjoy the light shining on the tiles at the top of the staircase.

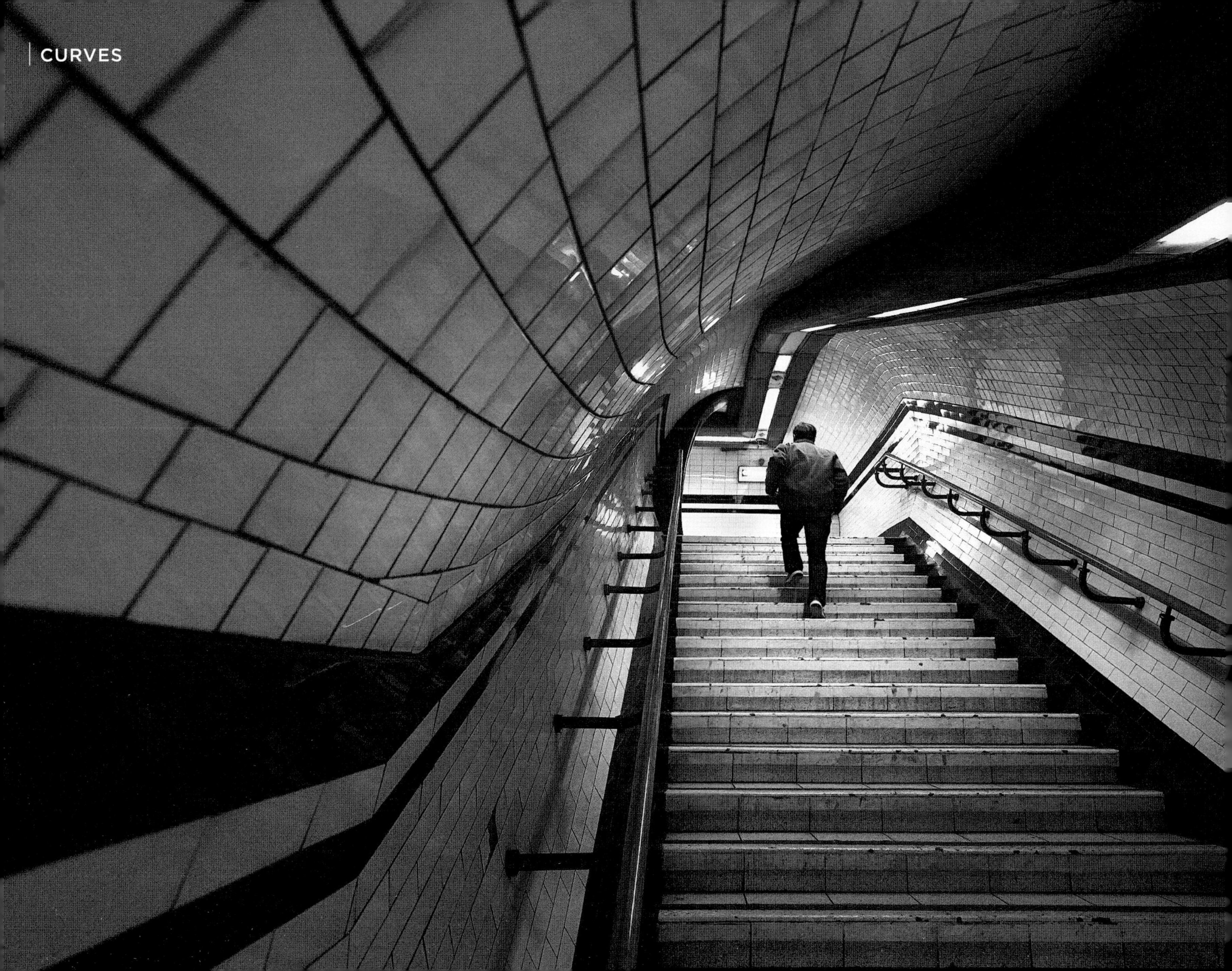
CURVES

The Waiting Room – East Finchley
Sometimes you have to wait a while to capture the perfect photo. I couldn't believe my luck when I happened upon two people sitting in nearly symmetrical composition while reading their newspapers; it was a very fortunate encounter.

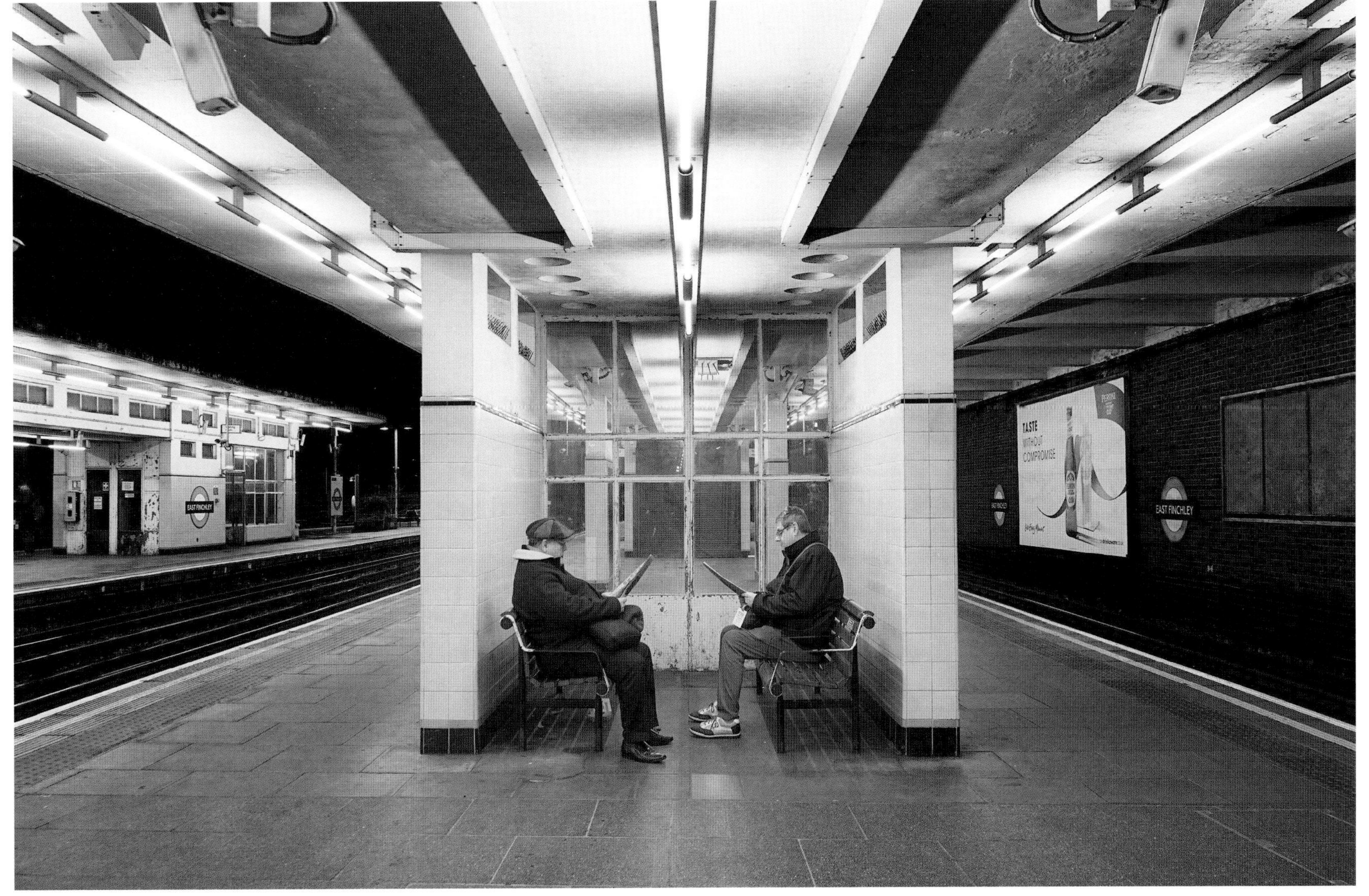
EAST FINCHLEY
TASTE WITHOUT COMPROMISE
EAST FINCHLEY

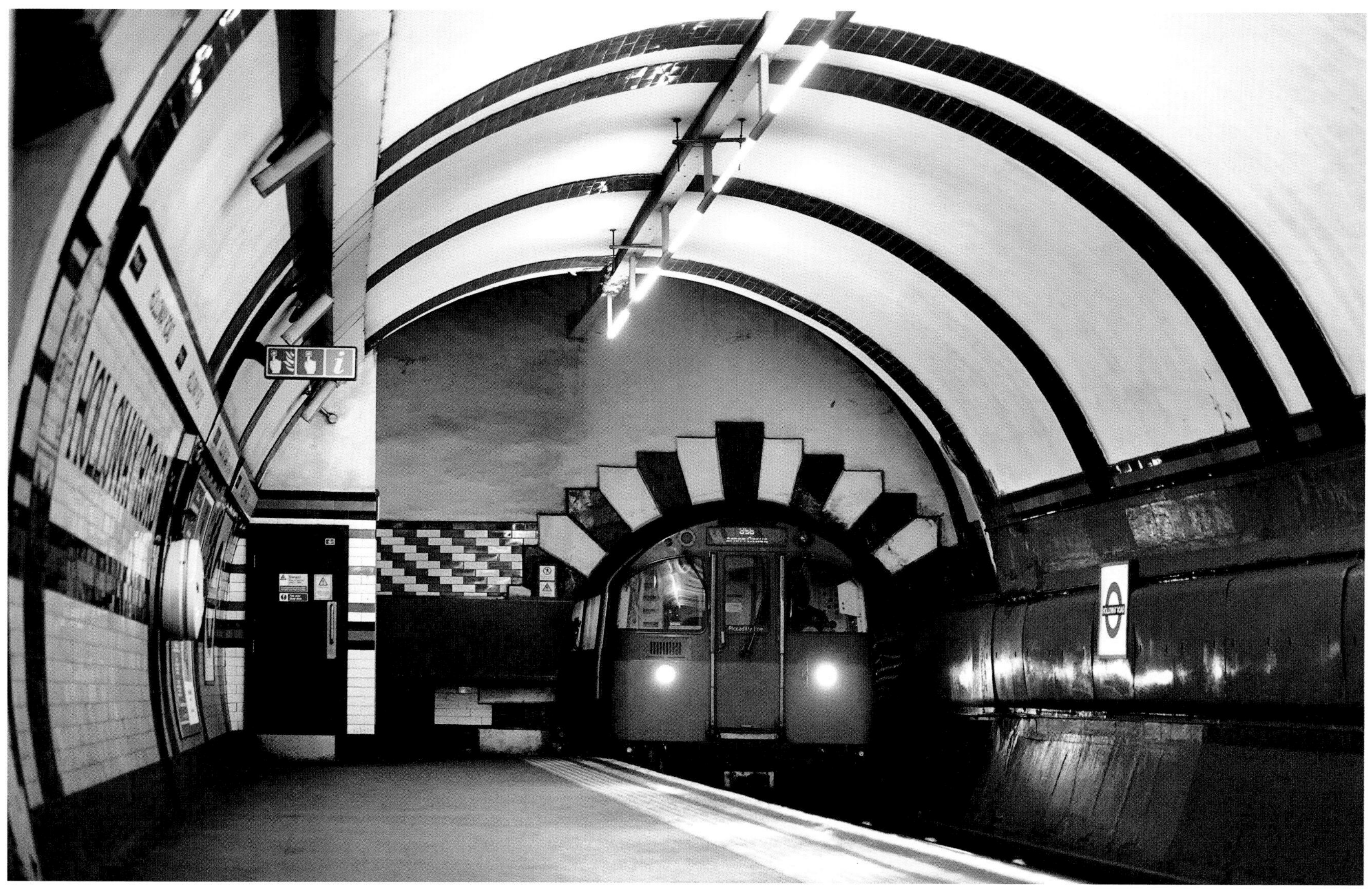

Tunnels, Trains, Twists and Turns – Holloway Road
I love how nicely decorated some London Underground stations are, like the curves, patterns and textures on display at Holloway Road.

5

LONG EXPOSURE/SHORT EXPOSURE

As we stand on station platforms waiting for our trains to arrive, our minds focus on the passing of time. Seconds, minutes and hours rush by, like trains zooming down a platform. The camera, however, perceives time differently from humans. Cameras cannot translate units of time into a still image; they capture moments. These can be short or prolonged.

Short exposures freeze time, capturing fleeting instances, such as stopping a bird in motion or slowing down a speeding vehicle. Long exposures display the passing of time, visually conveying any changes or movements that occur within the duration of the exposure, like a blurry train travelling down a platform for half a second.

Here's a collection of photos exploring long- and short-exposure compositions.

Photo Hints

Long exposure: Most of my long-exposure photos are taken handheld. To help get steady shots, adopt a balanced stance or lean on a wall, shoot in burst mode, and, if you have it, use a camera with image stabilisation in the camera or lens.

Short exposure: You must adjust your camera settings depending on your subject. Speeds over 1/250 will freeze moving people, and over 1/2500 stop in-motion animals and fast-moving trains.

A Long Departure – Church Street
I captured this fun and experimental ten-second exposure of a tram slowly departing the Church Street tram stop on a dark evening.

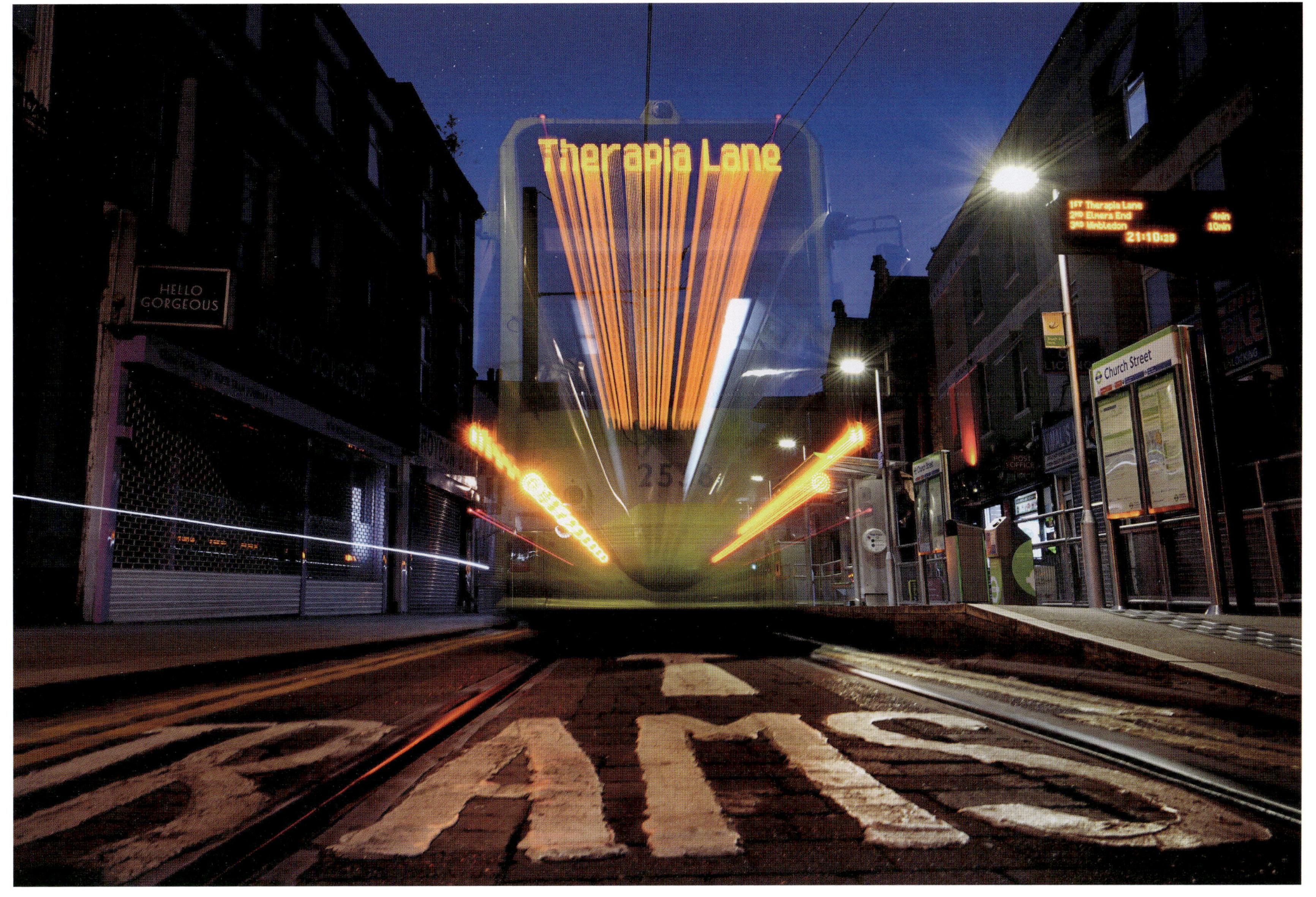
Therapia Lane
HELLO GORGEOUS
1ST Therapia Lane   4min
2ND Elmers End   10min
3RD Wimbledon   21:10:28
Church Street
POST OFFICE

A Bird Looks for Heron Quays
I wish I'd captured a photo of a heron outside Heron Quays station for my photography pun series, but I'm still pleased with this shot of a European herring gull looking for food near West India Quay DLR station.

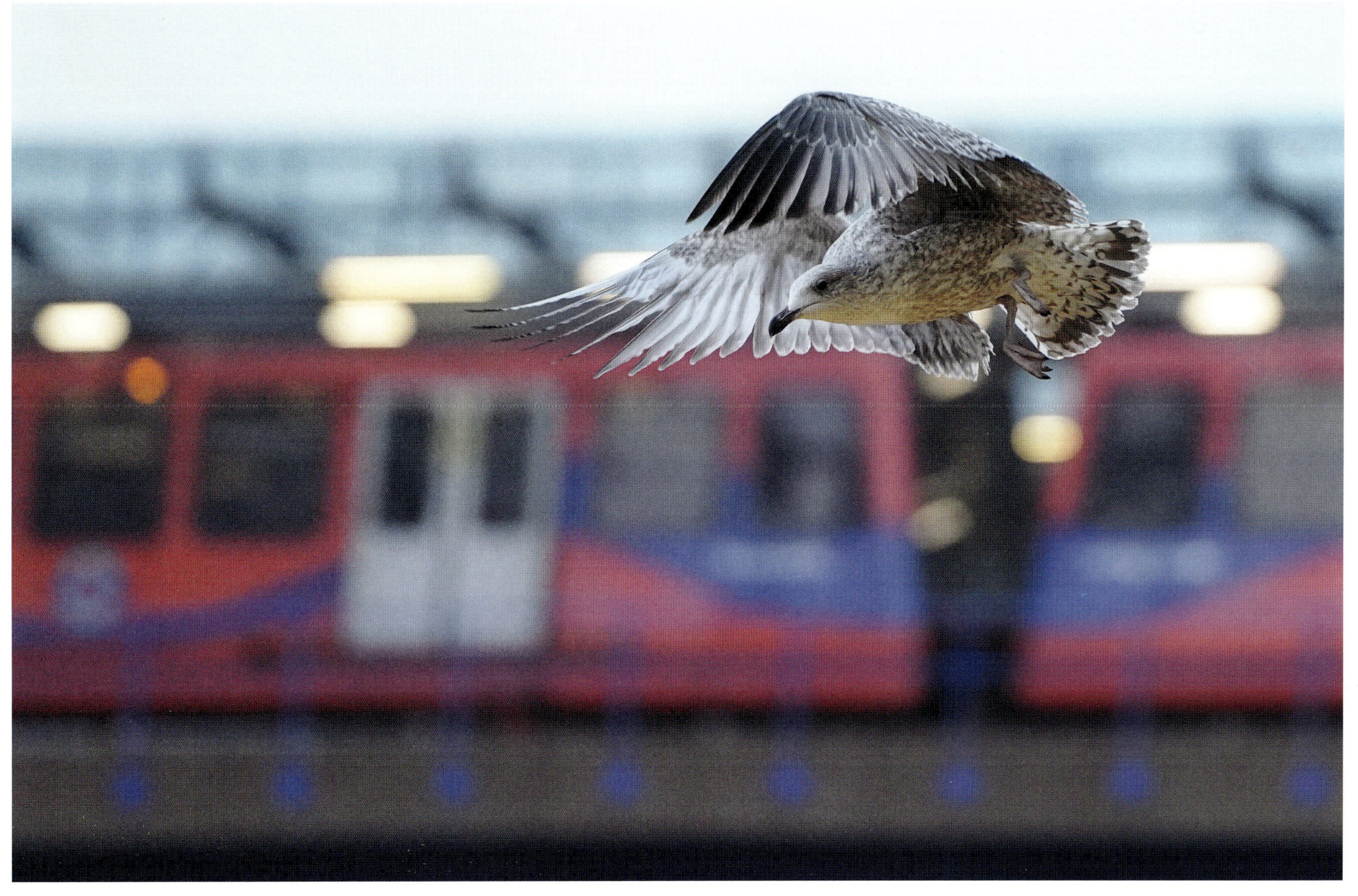

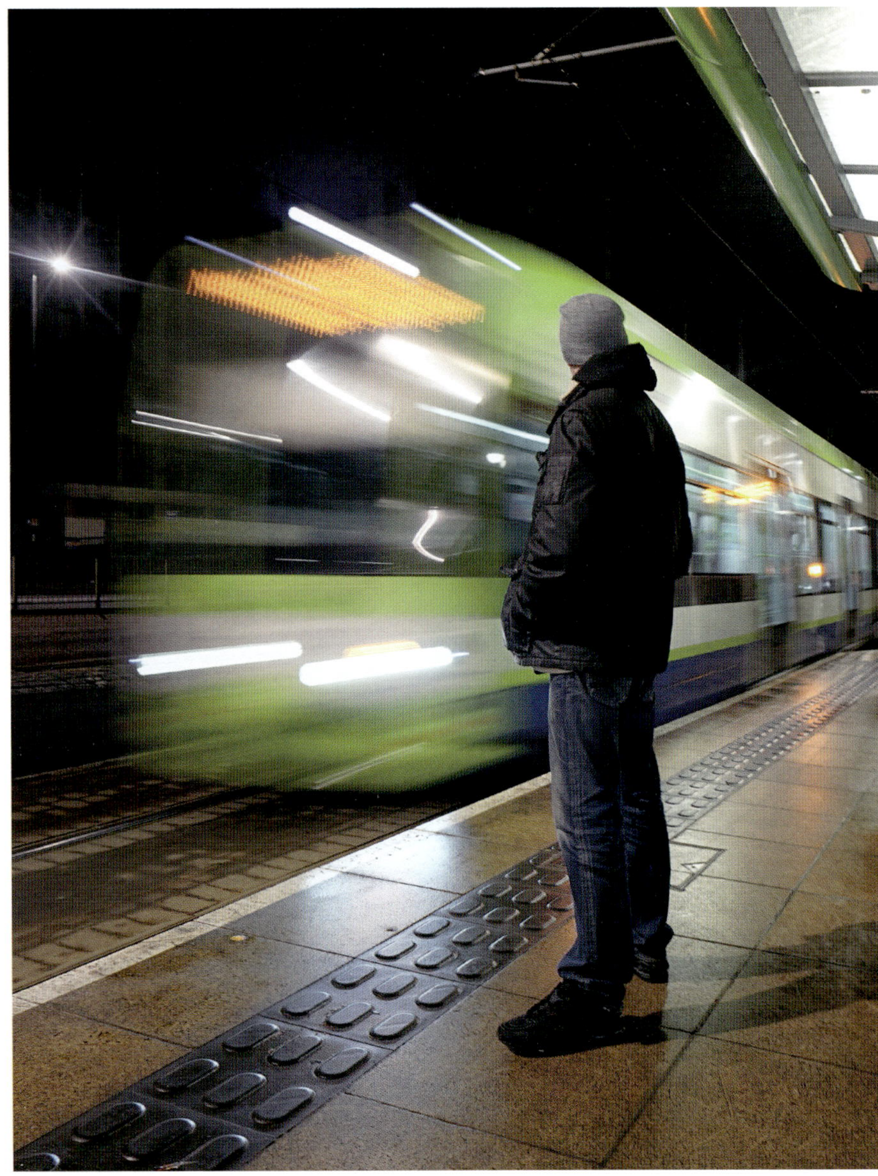

Waiting for Trams – Wellesley Road
Handheld long-exposure photography of a traveller waiting for a tram
to arrive at the Wellesley Road tram stop. I like the contrast of the bright
orange, yellow and green against the night sky.

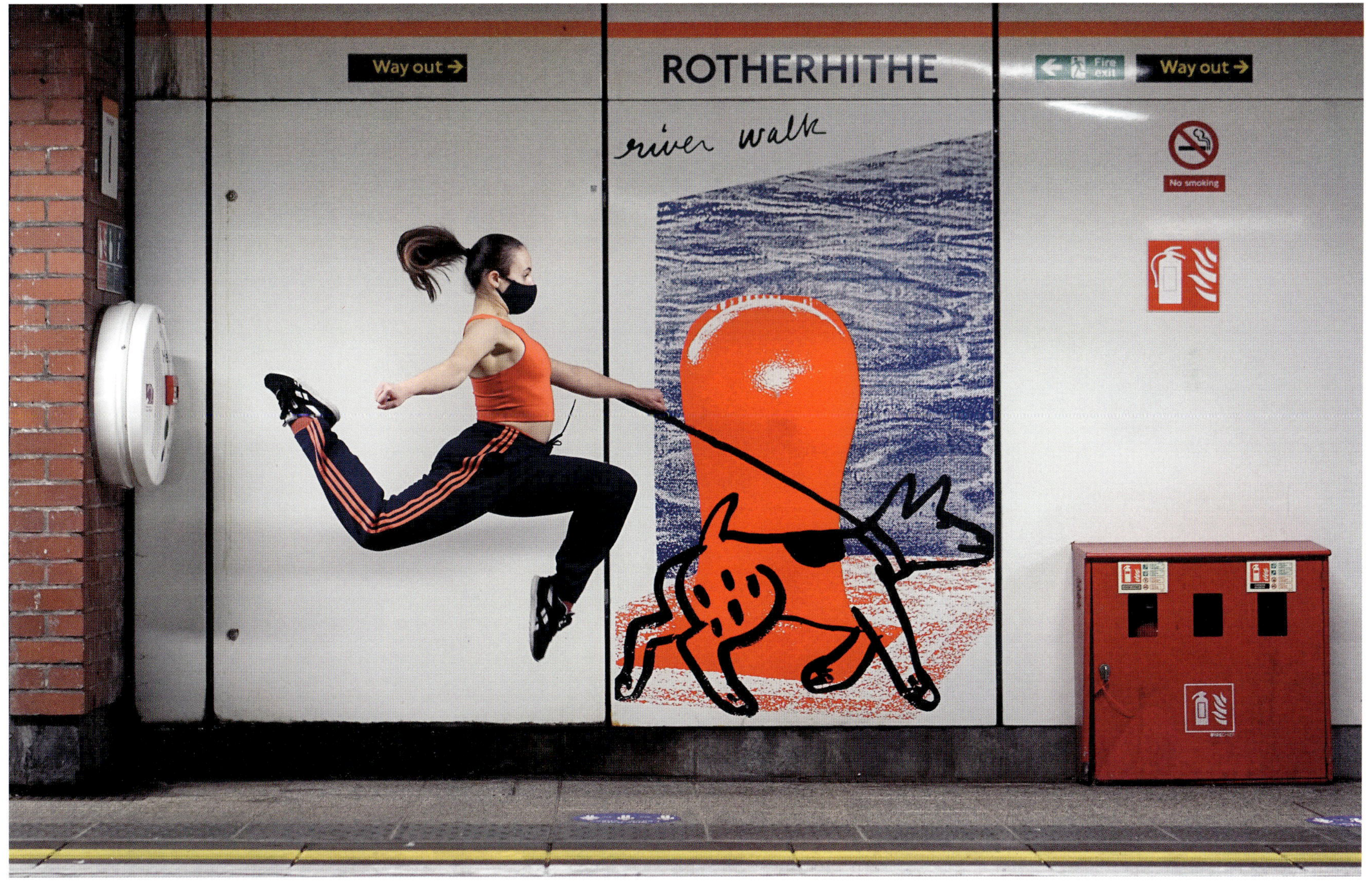

Walking the Dog – Rotherhithe
Andreya pretends to walk a dog as she interacts with the mural by Sue Huntley and Donna Muir at Rotherhithe Overground station.

Zooming Along the Escalator – Bank
Just like a perfect sunset, this particular advert, wrapped around the travelator at Bank station, was ideal for a long, colourful exposure. Andreya gracefully held a pose to complement the composition.

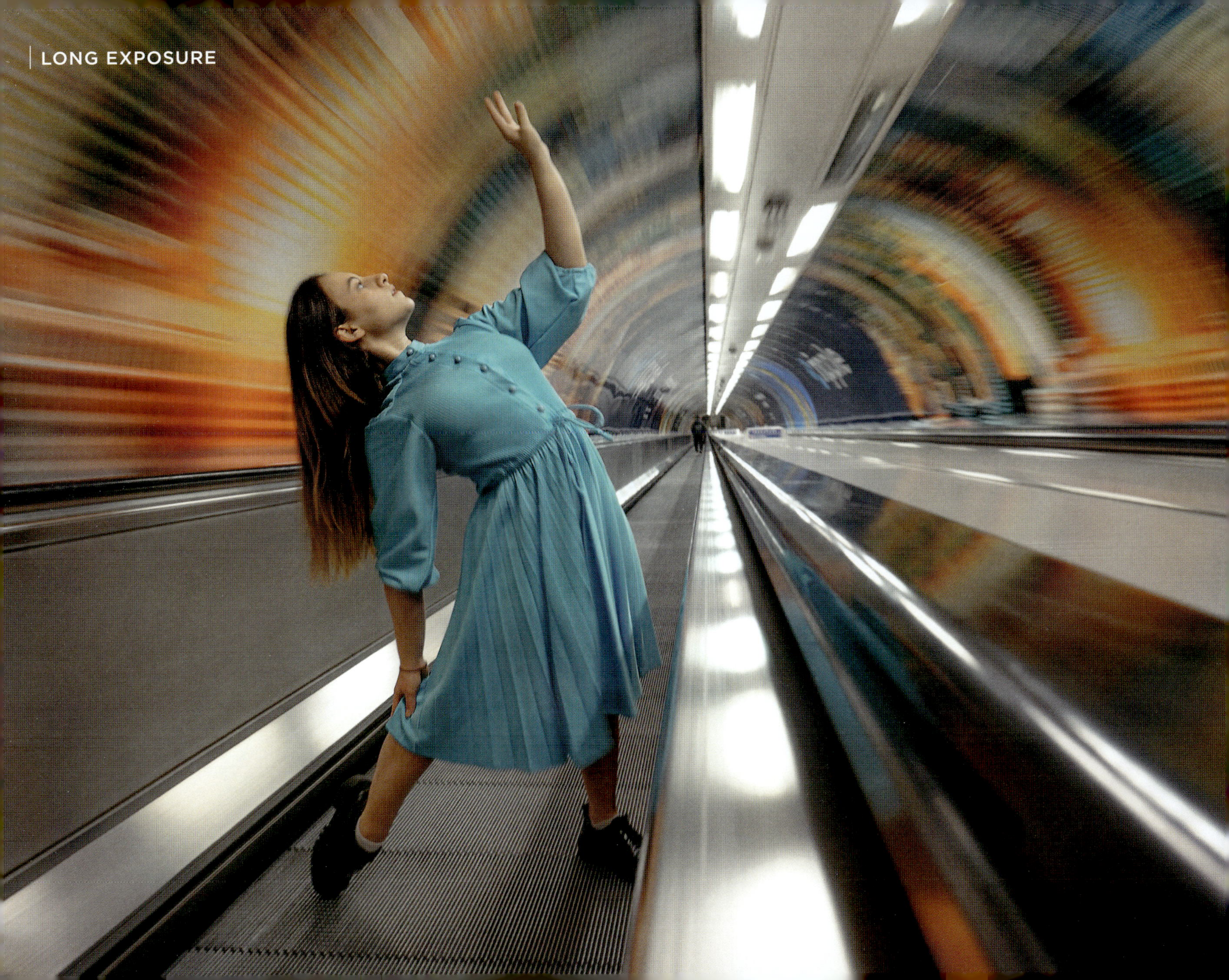
LONG EXPOSURE

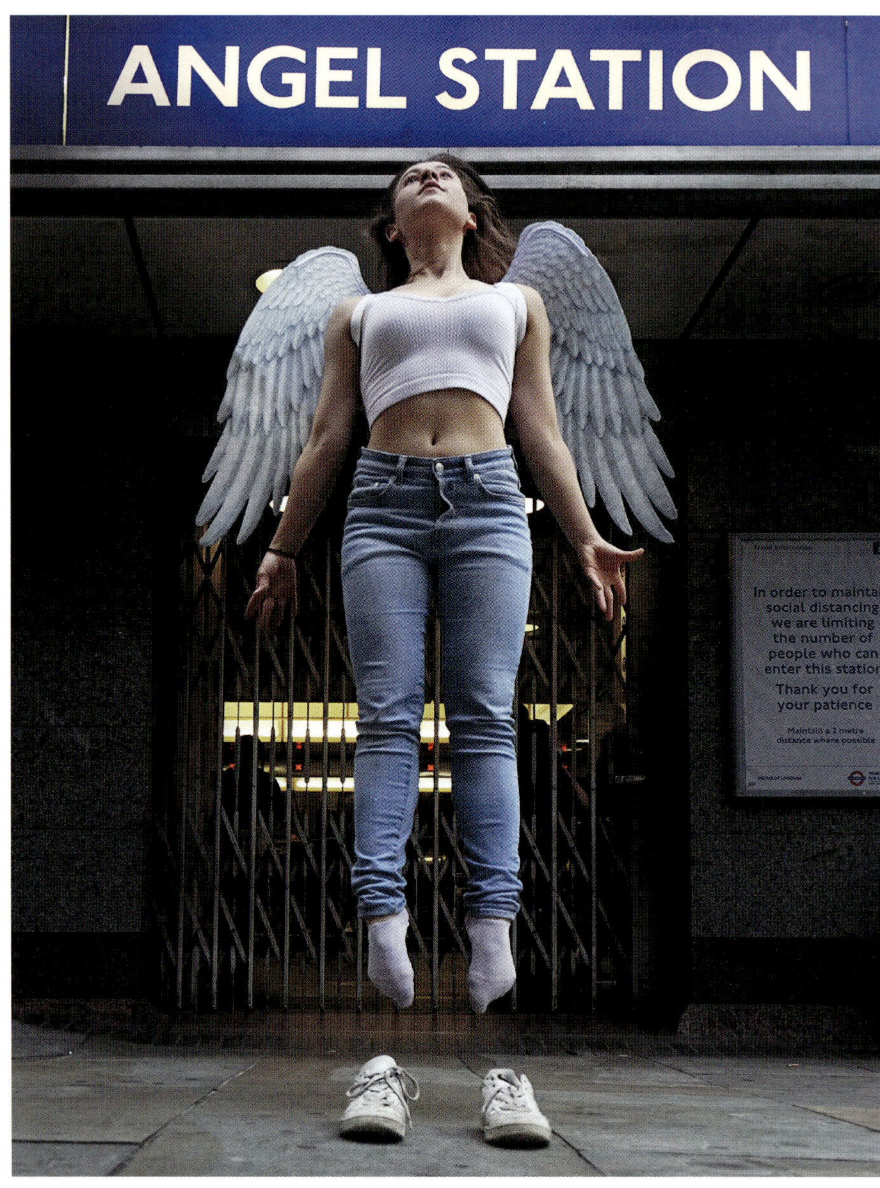

Light on Her Feet – Angel
Andreya wears angel wings outside Angel station, which is named
for the Angel Inn, a nearby seventeenth-century coaching inn.

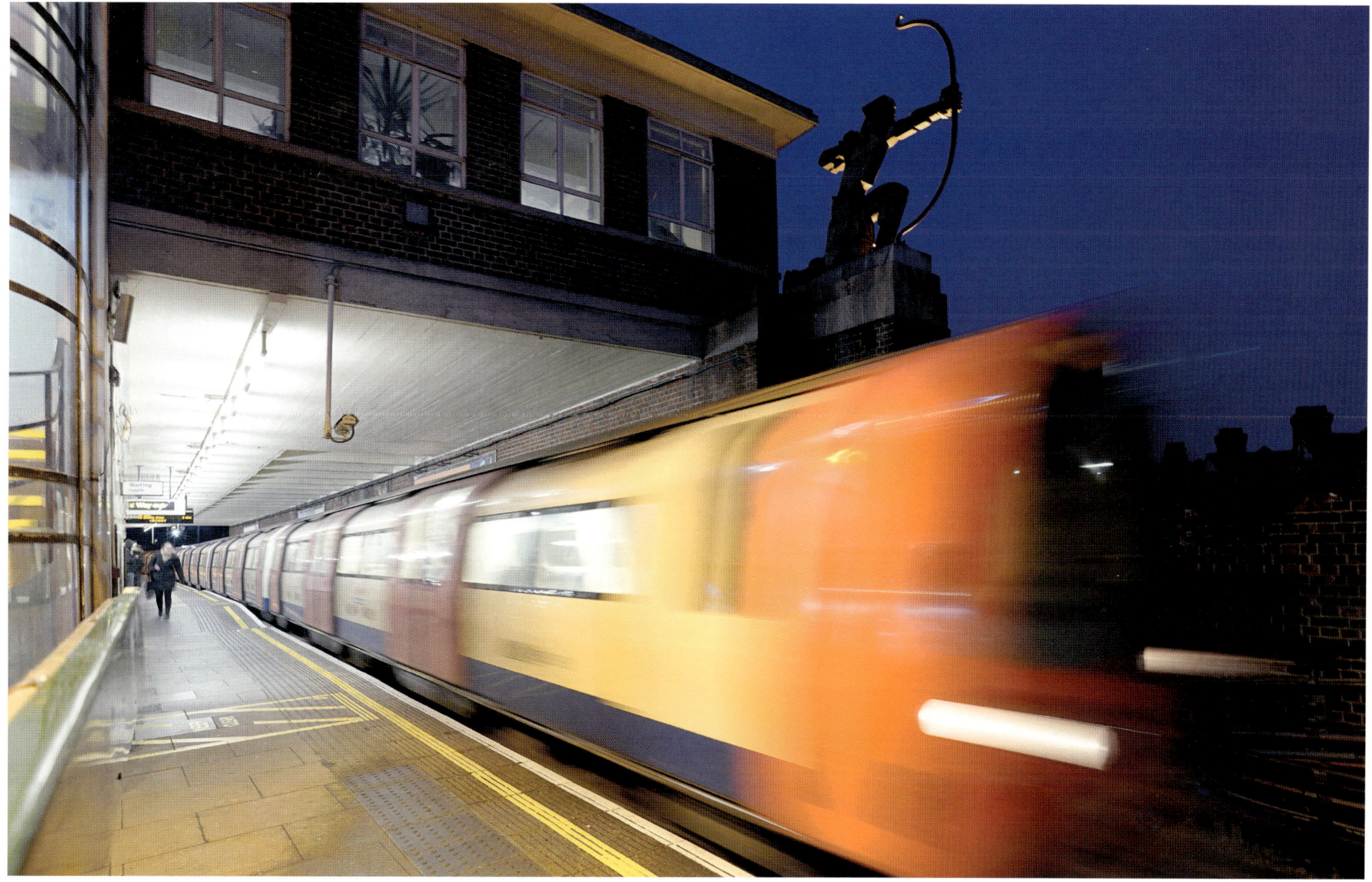

Shooting Down the Track – East Finchley
A handheld long exposure features the East Finchley Archer, or 'Archie', overlooking a train zooming down the platform.

Colourful Dance – Canary Wharf
Adam Nathaniel Furman created a permanent rainbow design on the supports of Adams Plaza Bridge, which is next to the entrance of the Canary Wharf Elizabeth line station. Here's a photo of Andreya dancing outside.

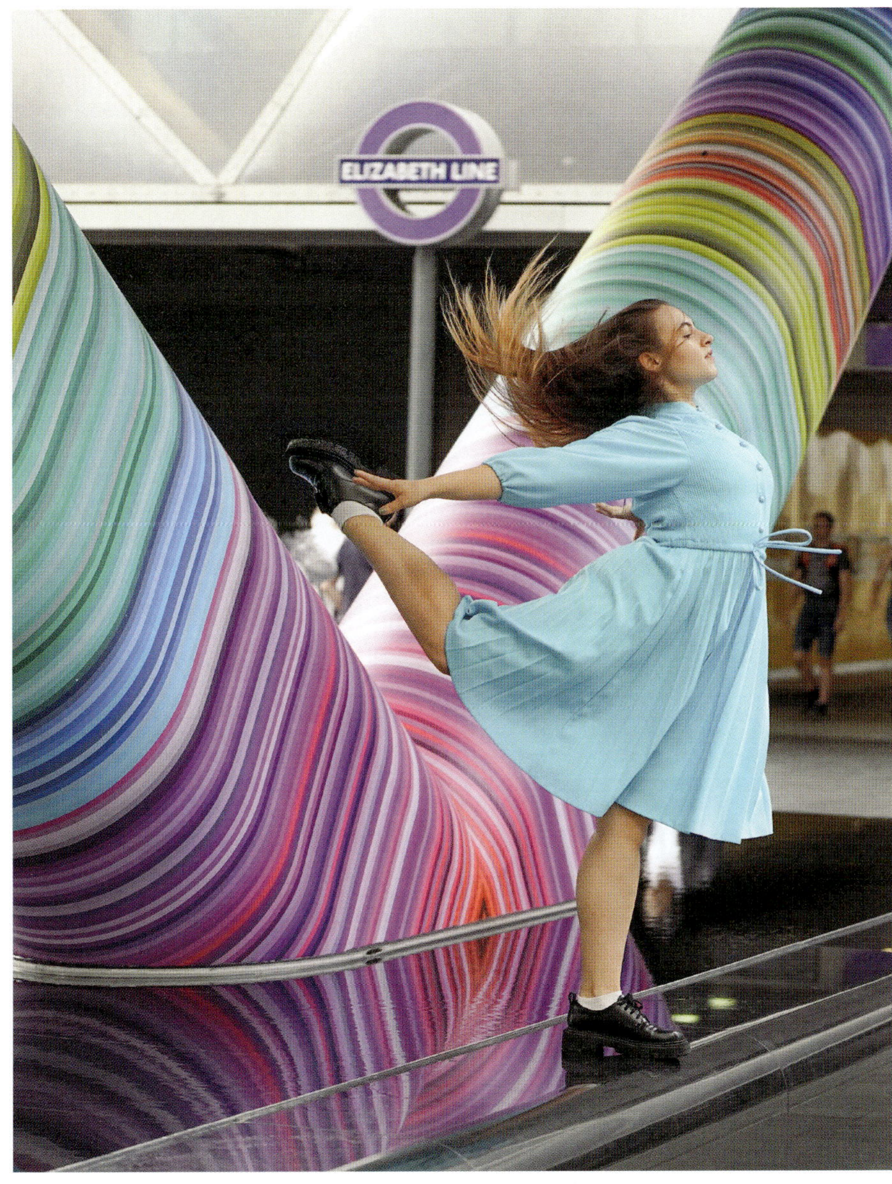
ELIZABETH LINE

Turbo Speed – Cutty Sark DLR

If you're lucky, you might win the front-seat position when you board the DLR; not only can you pretend to drive the train, but it also gives you a chance to capture photography like this 0.3-second exposure.

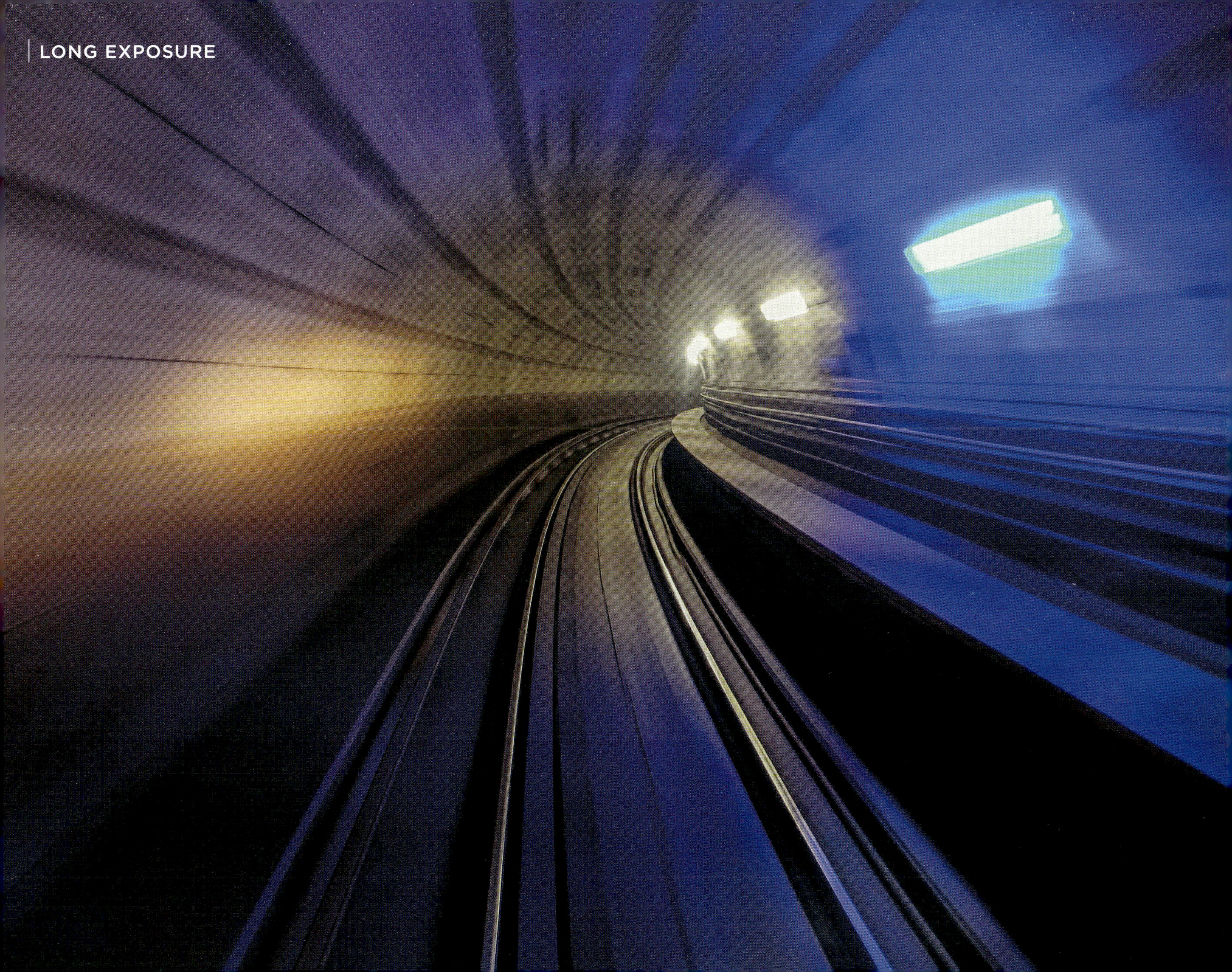
LONG EXPOSURE

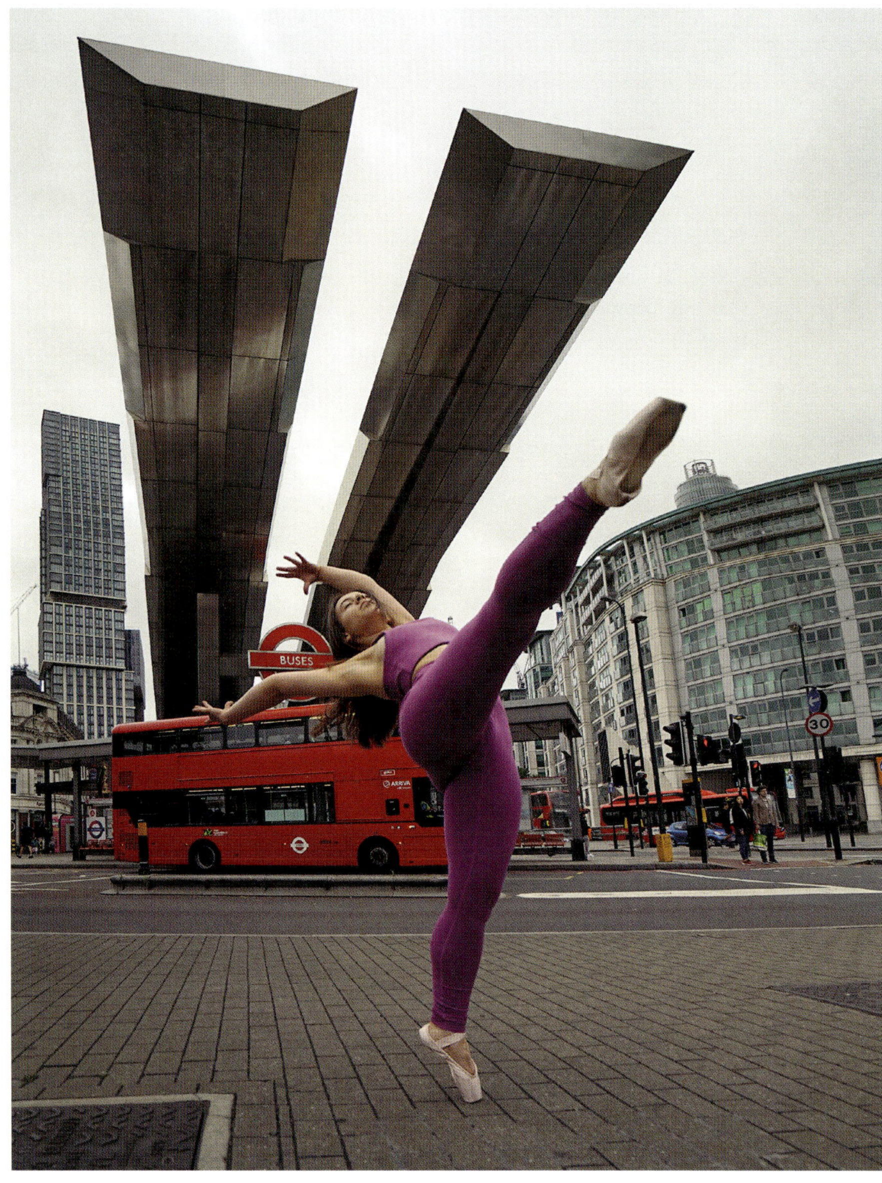

Pointe at Buses – Vauxhall
Andreya embodies the drama of the striking architecture of Vauxhall station and matches some angles with this kick pose.

Mythical Lines

Down below
the tunnel screeches
Cerberus growling.
High above
Hermes rides the rail
with weary commuters.
Both in a parallel dance
keeping the urban flow
steady and sure.

Danny Herbert

Tooting – Tooting Bec
Singer and musician Fiona Fey pretending to toot her clarinet at Tooting Bec Tube station.
Amusingly, 'bec' is the French word for the clarinet's mouthpiece.

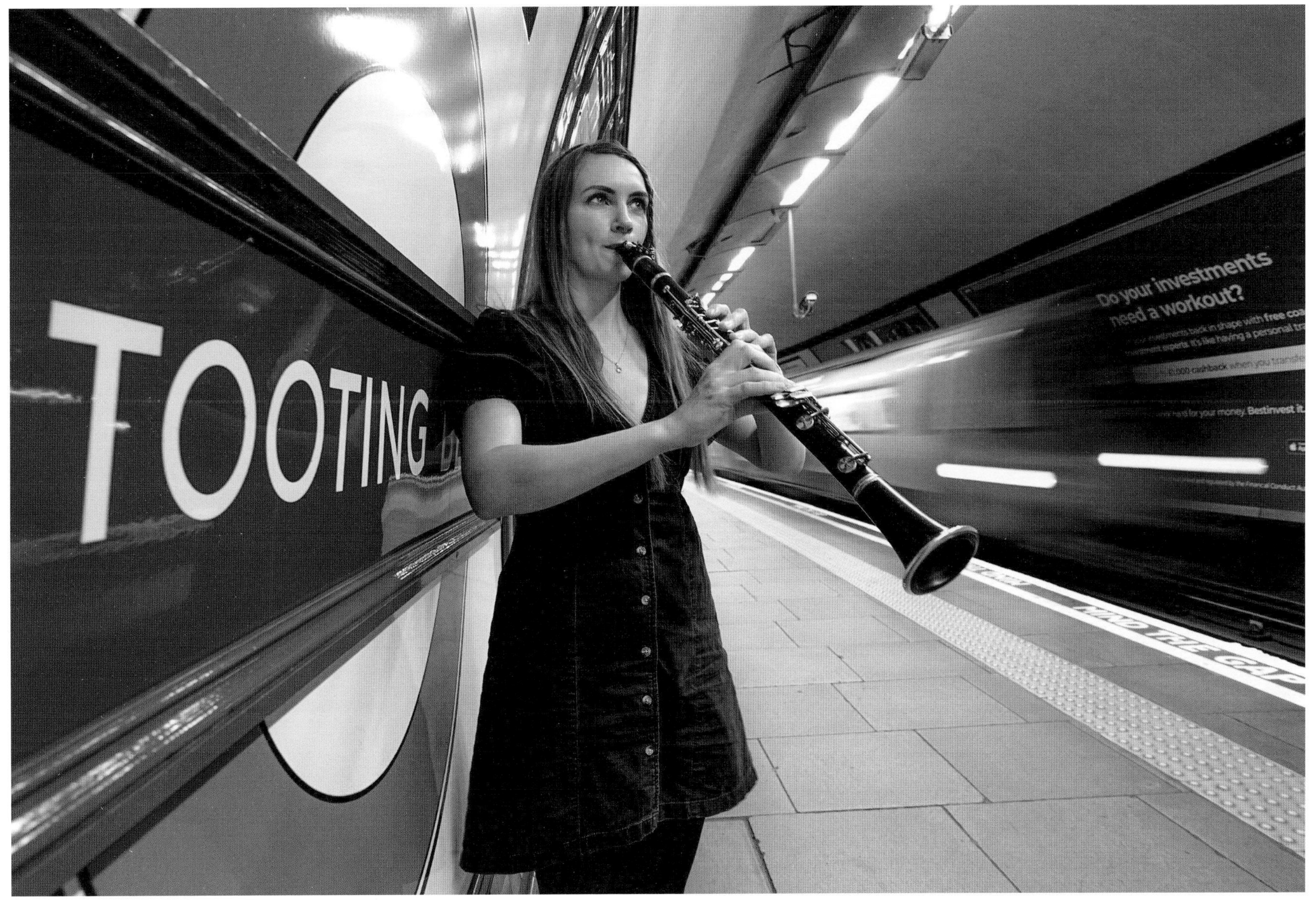
TOOTING
Do your investments
need a workout?

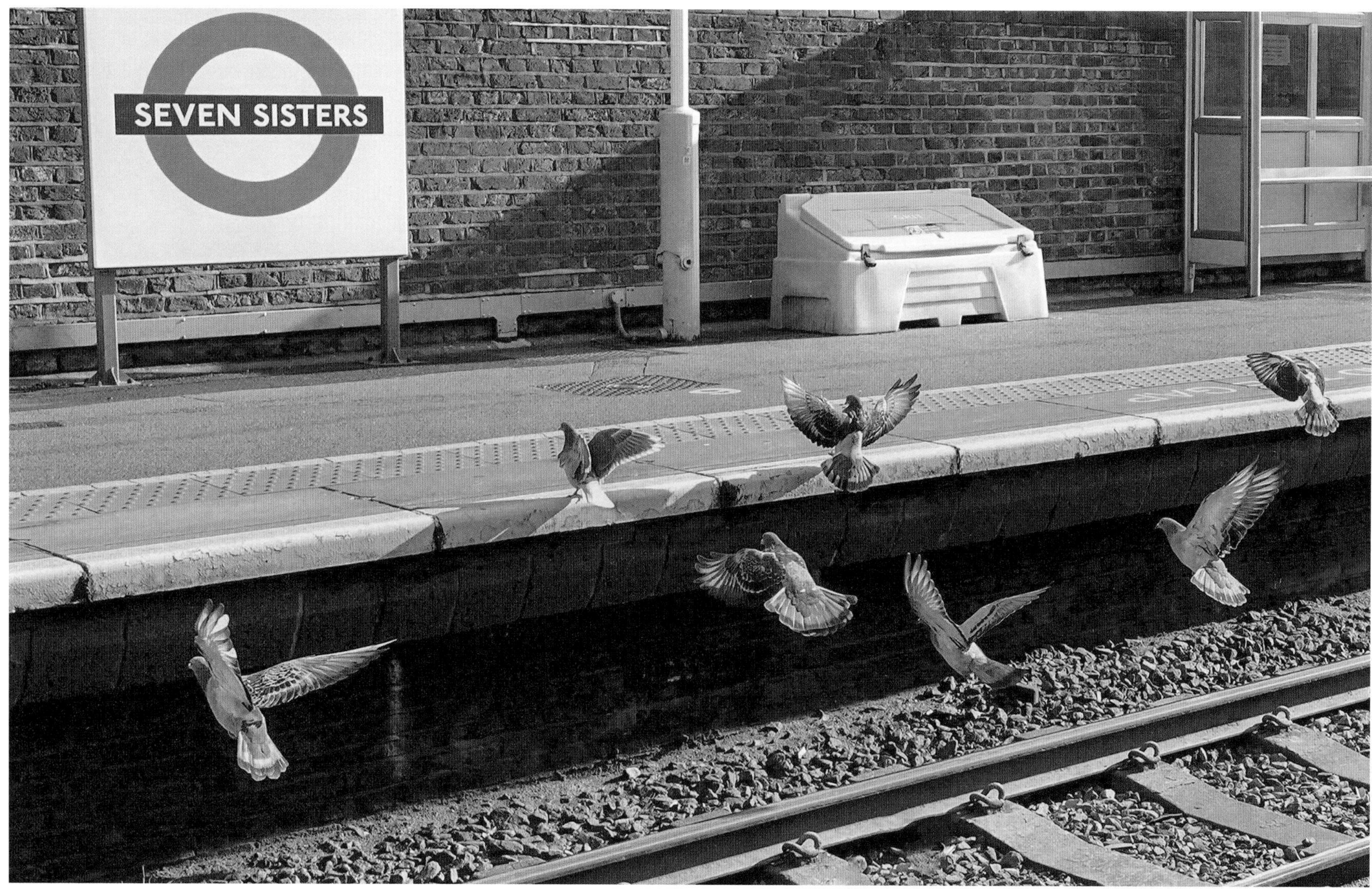

Seven Birds for Seven Sisters
In a serendipitous encounter, I photographed seven pigeons at Seven Sisters station. I'd like to think that they were all siblings ... but they probably weren't.

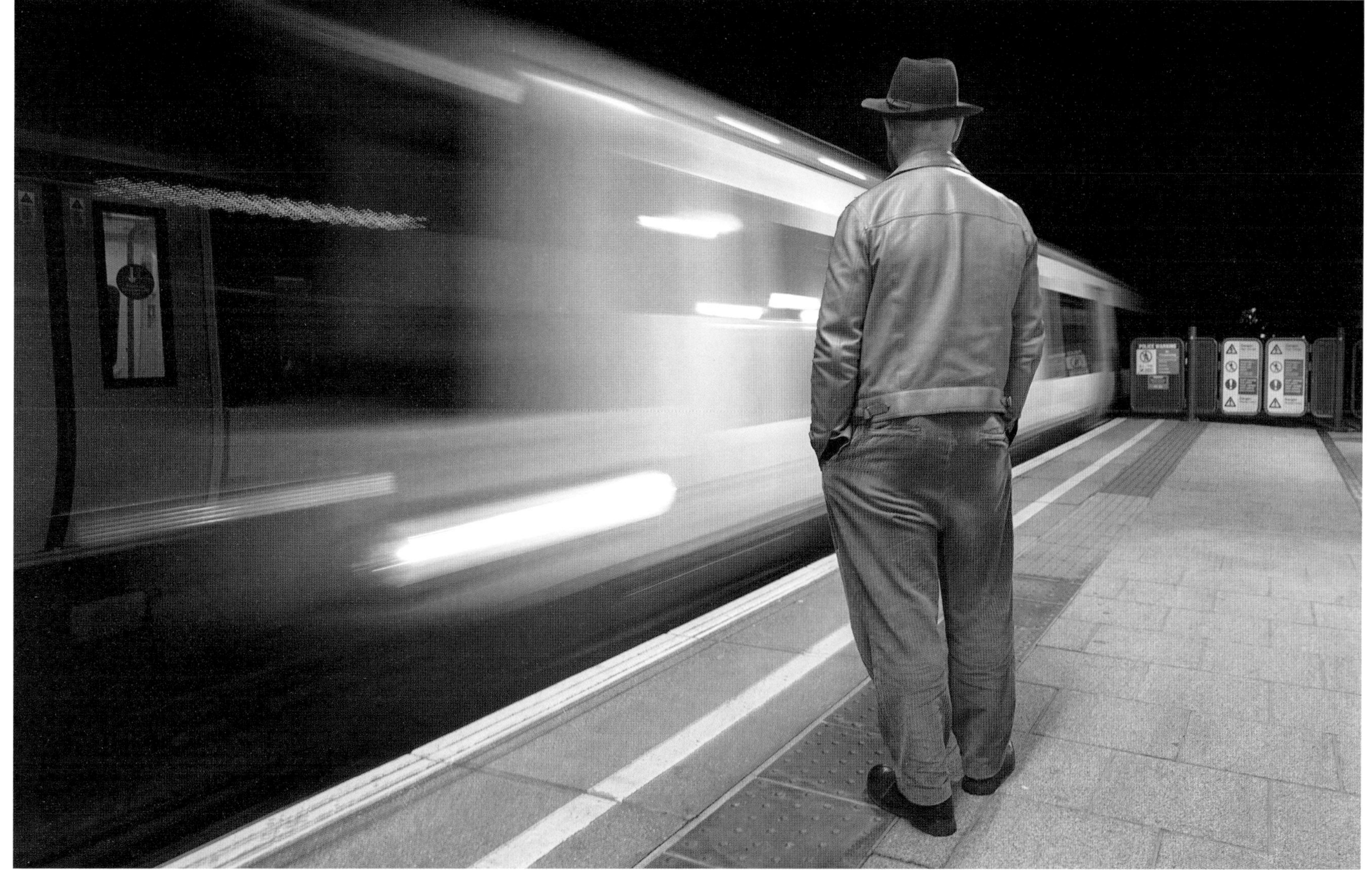

Indiana Jones and the Riders of the Lost Train – Haggerston
I always get excited when I see a person wearing a hat; I have a growing photographic collection of hat-wearing people on the London Underground network.

Birds of a Feather – Marble Arch

Pigeons are regular users of the London Underground, and I always enjoy capturing these peculiar commuters using the transport system. These guys were polite and posed on top of the station entrance signage.

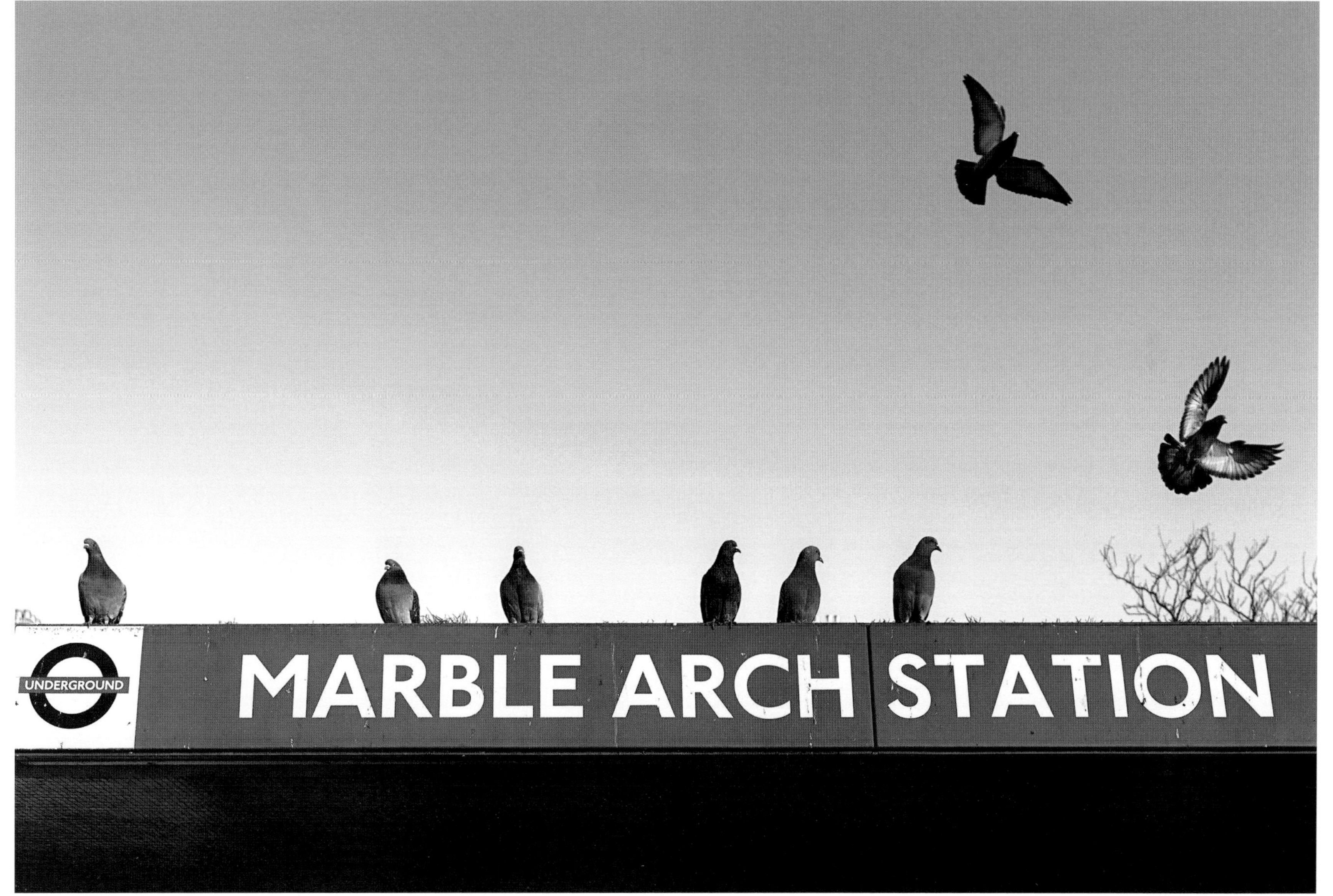
UNDERGROUND
MARBLE ARCH STATION

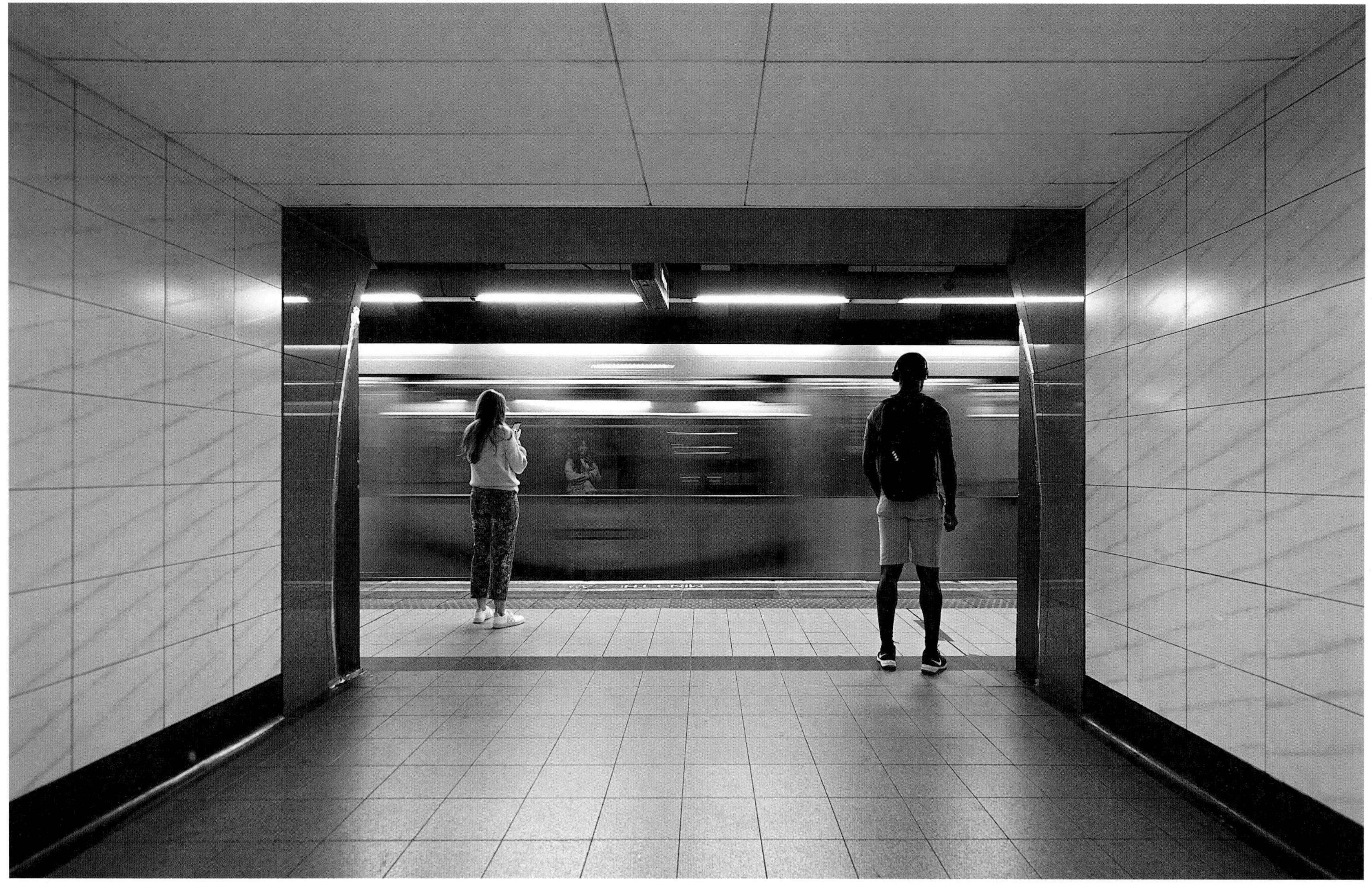

Waiting & Reflecting – Bank
Two commuters wait patiently for their train to arrive. I love the reflection of the lady in the blurred moving window.

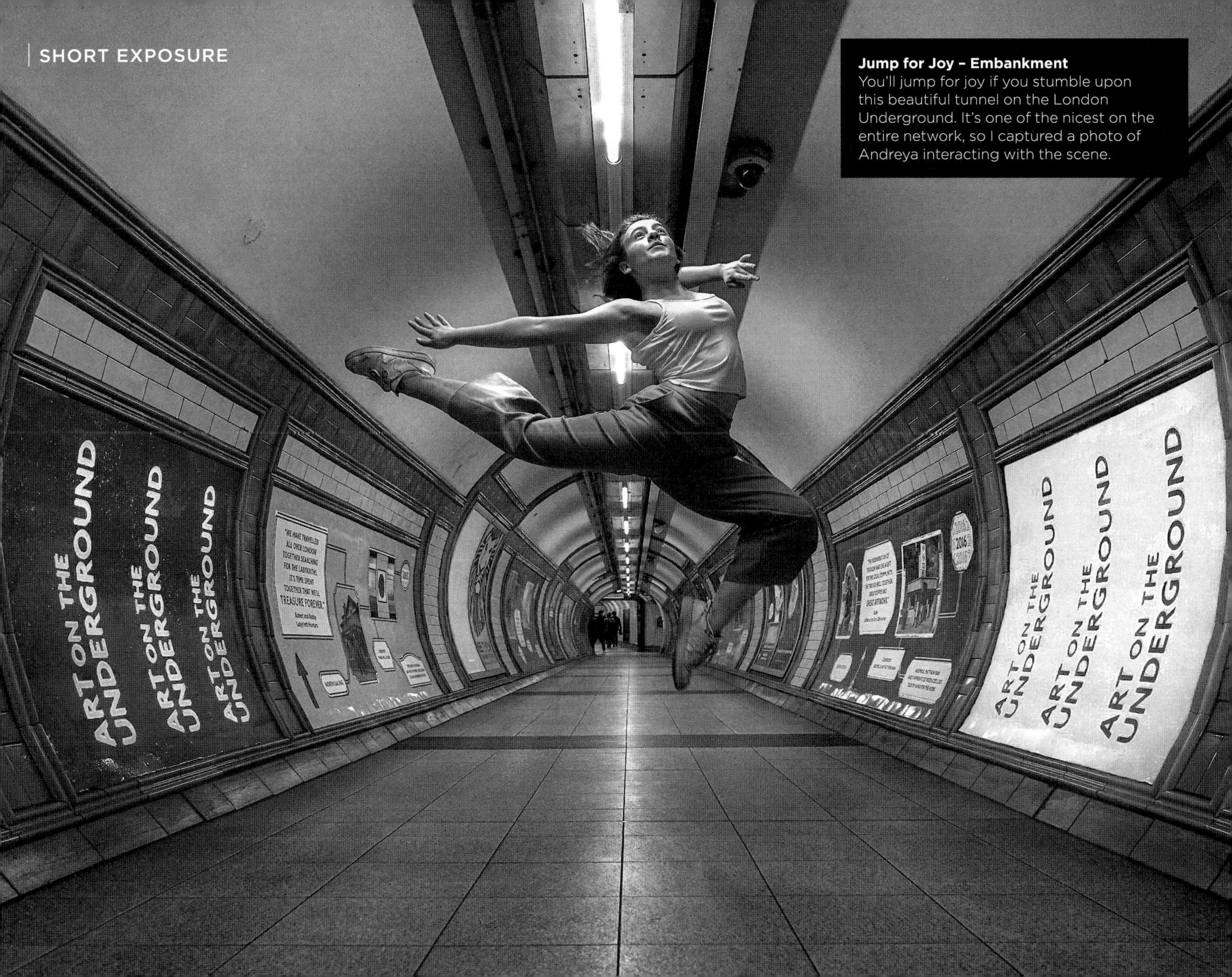

Jump for Joy – Embankment
You'll jump for joy if you stumble upon this beautiful tunnel on the London Underground. It's one of the nicest on the entire network, so I captured a photo of Andreya interacting with the scene.

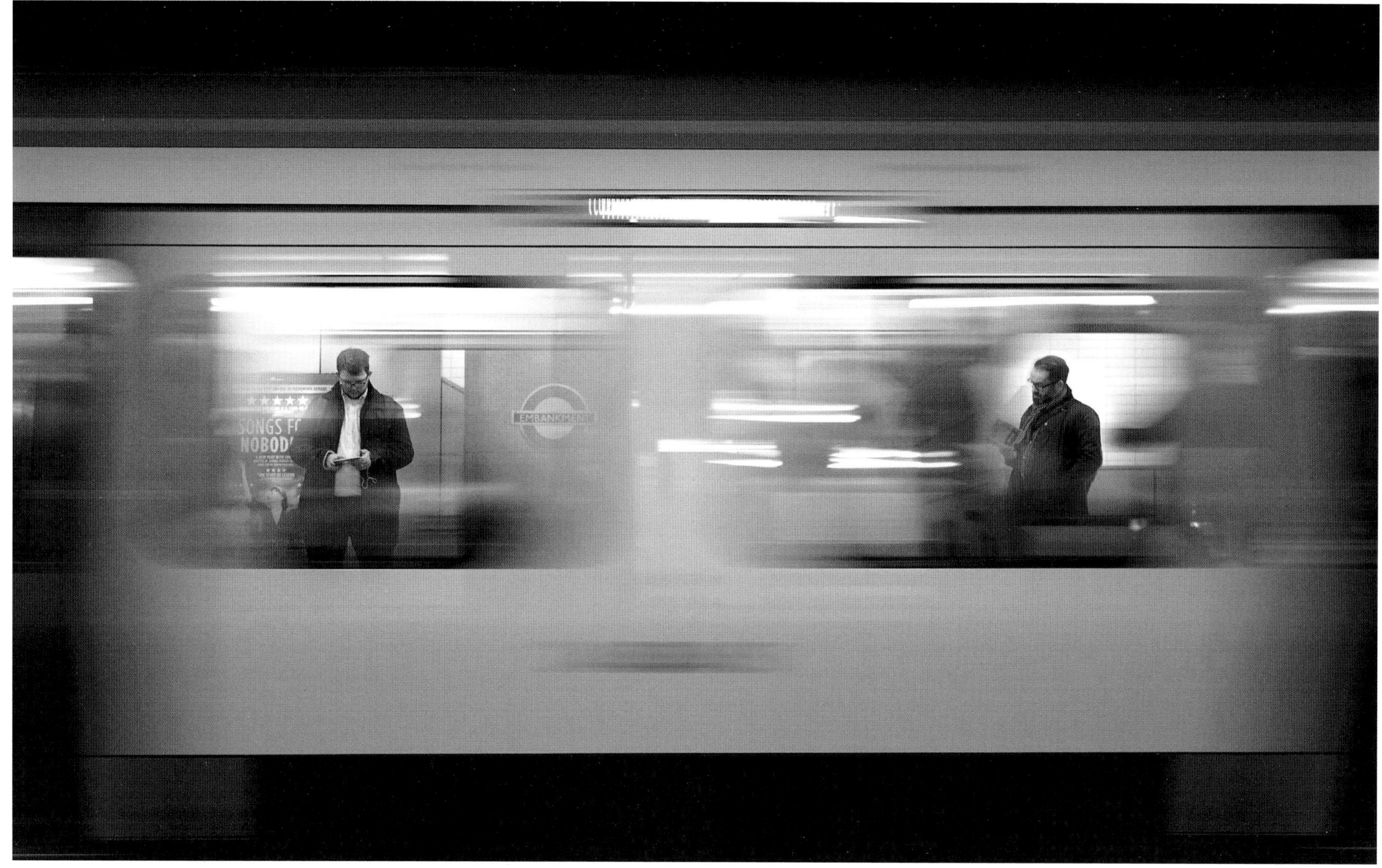

Through the Looking Glass – Embankment
I love capturing photos of people waiting while looking through the moving windows of a train.
The shot requires practice and plenty of luck, but will be worth all the effort.

CAMERA AND PHOTO-EDITING INFORMATION

Here is a quick summary of my choices for editing, cameras and lenses. For those interested in a more in-depth guide, I have provided a link to one on my website.

Photo Editing

I edited my photography using Photoshop. As always, I do as little post-work as possible and mainly just clean up, removing the standout objects and random limbs as people rush through stations. A few of the photos taken are composites or in-camera double exposures.

My Camera

My images are composed quickly and are often taken in tight spaces. Because of this, my main requirements for the project are a wide-angle lens and a camera body with a high megapixel count. These features allow me to crop speedily taken compositions. Since all shots are handheld, another necessity is image stabilisation.

Because of my needs, the camera I use is a Sony A7riii. I use a 16–35mm f4 lens for ultra-wide shots; a 20mm 1.8 G for wide shots when there is a lack of light or when I need greater control of the depth of field; and a 55mm 1.8 for more zoomed-in compositions. I've also recently added a 70–180mm to my collection to capture distant objects.

Photography Guides

As I am frequently asked about how I capture images, I have written some free online guides that explain my methods. These can be found at **tubemapper.com/hello.**

The Journey Continues Online ...

If you want to see more, please visit me online, where you can view all the photos that I have captured so far. I document the project frequently, so this is the best way to keep up to date.

I also welcome you to submit your suggestions for the best views of the Underground, ask about photo techniques, or just send a message saying hello. Also, if you've enjoyed the book and have a spare moment, please leave a review online.

For more information about the Tube Mapper project, please visit **tubemapper.com.**

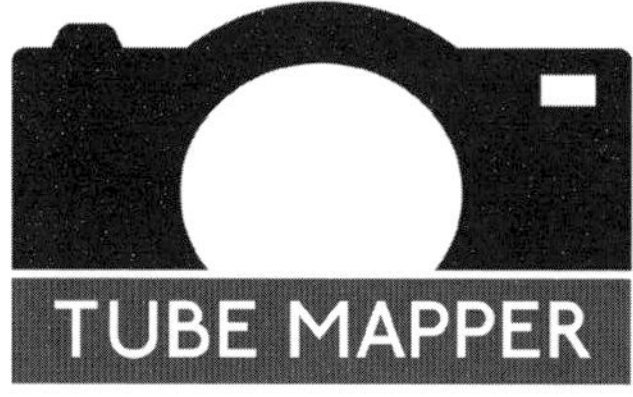

ACKNOWLEDGEMENTS

The London Underground remains a constant muse, delivering an almost endless supply of inspirational sights and experiences to share. It's truly an honour to be able to create these books and share them with you all. So I would like to acknowledge all of the help that I've had on this journey.

I thank my children and my amazing wife for their patience, support, guidance and unwavering faith in my crazy ideas. They may be used to my writing now, but it is always tricky and involves a lot of compromise and reorganisation. Thank you for believing in me.

I thank the writers Danny, Tracy, John, Victor, Michelle, and Matthew for their beautiful contributions to the book. You guys are eternally awesome.

I want to thank dancer Andreya and gymnast Anna for their creativity and ability to translate my ideas into art.

To all my family and friends, thanks for all the support and encouragement you gave me on this journey. Thank you, Mike, for your exceptional knowledge and help. Again, an extra special thank you to Edward for all your time and help.

I want to thank all the wonderful people I have met during the project, especially those online, for their support and for helping the project reach this destination.

Lastly, as always, thank you to Transport for London. I dedicate this book to all the workers, on the front line and behind the scenes, who help make the London Underground the most incredible transport network in the world.

With special thanks to the following for their support:

Marcus and Beatrice Agbaimoni
Daniel Andrews
Alan Appleton
Karen Arrighi
Mike Ashworth
William Baggley
Martin Baines
Louis Baker
Jean Barclay
John Barker
Martin Barnes
Christian Bast
Peter Beecham
Louis Berk
Marie Christine Berthommier
Trevor Blofeld
Andy Blurton
Patrick Bohlmann
Mark Bowman
Harry Brayford
Andrew Brown
Chris Brown
Joe Brown
Jenny Carpenter
Sharon Carneiro
James Chapman
Lisa Chudleigh
Robert Coates

Howard Cohen
David Collison
Jon Cooper
Stephen Cooper
Don Constance
Stephen Cotton
Mathew Coutard
John Cripps
Israel Crowe
Ben da Costa
Jasper Dade
Christopher Damoune
Lewis Day
Harry Dellow
Martin Dewar
Andrew Donnelly
Brian Doyle
David Easton
Samantha Eggleton
Christina Eneroth
Dave Field
Feilo
Kathleen Fleming
John Francis
Graham Foulds
Kiyomi Fujii
Malvin Fuller
Michael Fuller

Gordon Fullerton
Rob Gahan
Ashley Garrison-Brown
Nigel Gibson
Jonathan Goode
Adam Gordon
Peter Goulborn
Che Grant-Ford
Sarah-Jane Green
David Griffiths
Ian Griffiths
Justin Grimwade
Nicky Hall
Elizabeth Harding
Matt Hayes
Russell Heath
Dave Heath
James Hendry
Matt Higgins
Paul Horner
Phil Hort
Jason Howat
Markus Hübner
David Ilott
Melissa Jade
Luke A. Johnson
Brendan Kavanagh
Rowan Keenan

Donald Kennedy
Kerry and Simon
Hans Kristian Henriksen
Alan Lathwell
Peter Livermore
Karl S. Leutner
David Lloyd
Locky and John
Jay Long
Nat Lowden
Keir McArthur
Ken McCallum
Richard McCarthy
Donald MacRae
Michael McIlhone
Jim McMullin
Ian Maddocks
Jonathan Marsh
Neil Maughan
Henry Mensch
Elizabeth Mills
James Mills
Jane Minton
Alistair Minton
Lakshay Mohan
Lindy Louise Mollison
Ruth Munro
Timmy Munro

Natalie
Thomas Newell
Jan Niemczyk
Paddy O'Hanlon
Peter Orr
Leo Papic
Robbie Perot
John Portch
Andrea Pomponio
Lindsey Porrett
David Pritchard
Alistair Proudfoot
Jeremy Rawlings
Joseph Roberts
Matt Roberts
Bjarni Rúnar Ingvarsson
Jonathan Russell
Nigel Russell-Sewell
Roman M. Russocki
Mark Sandland
Phil Shemmings
Adam Slatter
Mark Shone
Nicholas Shepherdson
Colin Smith
Susan Smith
Katie Spoo
Jacob Stevens

David Stone
Ian Stringfellow
Kristoffer Strömblad
Richard Stumpf
James Thomson
Colin Douglas Thornton
Colin Tompson
Jennifer Tillier
Nick Triviais
Roger Tuke
Emily Turner
Matthew Ward
Jonathan Warrington
Tim Wayne
Thomas Webster
Lewis and Glenis Weidman
Judith White
Stephen White
Paul Whittaker
William White
Suzi Williams
Kirsty Wilson
Stuart Wilson
Sean Wright
Simon Wyatt
Len Vinci
Aaron Ubasa

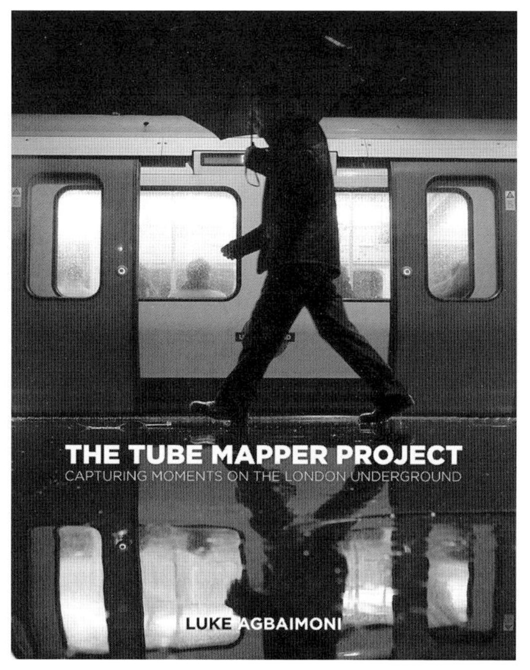

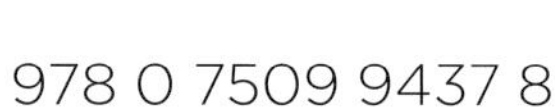

978 0 7509 9437 8

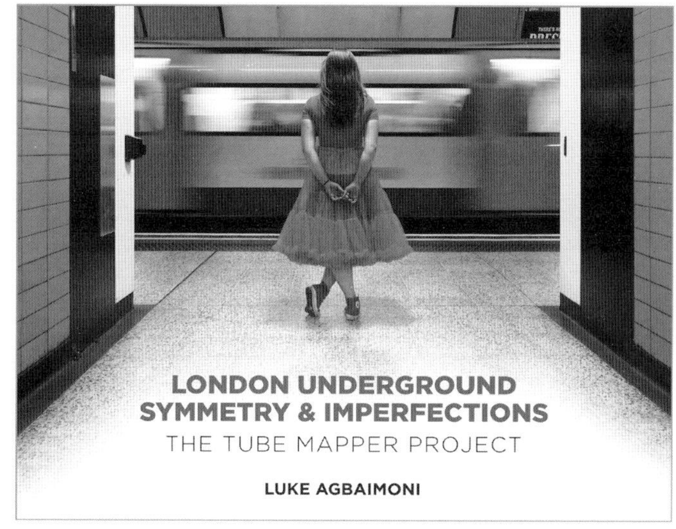

978 1 80399 157 3

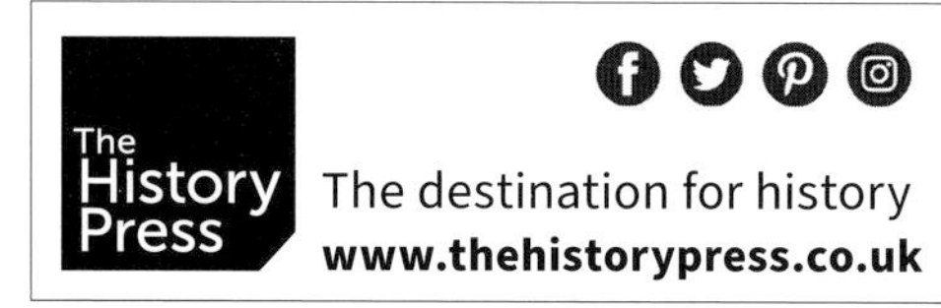